Dynamics of Aircraft Flight

Series Editor
Jean-Paul Bourrières

Dynamics of Aircraft Flight

Gilles Louis

iSTE

WILEY

First published 2022 in Great Britain and the United States by ISTE Ltd and John Wiley & Sons, Inc.

ISTE Ltd
27-37 St George's Road
London SW19 4EU
UK

www.iste.co.uk

John Wiley & Sons, Inc.
111 River Street
Hoboken, NJ 07030
USA

www.wiley.com

Library of Congress Control Number: 2022932439

British Library Cataloguing-in-Publication Data
A CIP record for this book is available from the British Library
ISBN 978-1-78630-719-4

Contents

Foreword

When Gilles invited me to preface his book on the mechanics of flight systems, I immediately accepted. My interest in his work is not only the result of our collaboration on multiple subjects and technical projects over the past 12 years but is also based on the way he has dealt with the subject.

This book is not simply a collection of mathematical formulas; it is the fruit of experience coupled with an ability to conceptualize. Following a general guideline, each step is described in great detail, with a number of concrete examples to which mathematical formulations are added. The latter are of course the fundamental elements leading to the sizing and optimization of the systems.

The mathematics used in this context is not accessible without prior understanding. It is therefore out of a pedagogical spirit that Gilles first takes up all the definitions of the mathematical elements: lift, drag, aerodynamic focus and so on.

In conclusion, this book offers a global vision of what flight mechanics is and it is therefore of great interest to all engineers and scientists who are called upon to work in this field.

Serge ZANINOTTI
February 2022

Introduction

What characterizes the mechanics of flight in relation to other branches of rational mechanics is that, among the various forces present, the aerodynamic forces intervene, and they themselves result from the relative movement of the aircraft in relation to the surrounding atmosphere. It is therefore essential to define the characteristics of the atmosphere.

The notion of a standard atmosphere meets this need. Air is considered an ideal gas, represented by the three state parameters (pressure, density and temperature), heavy and still.

Just as it is necessary to define the atmosphere in which the airplane is traveling, so it is necessary to be able to measure the velocity on board the aircraft, especially the velocity in relation to the air.

The velocity of an aircraft is often given by the Mach number. This value is the ratio at a specific point between the velocity of the flow V and the speed of sound at that point. The "aircraft" Mach number is the ratio M_0 between the aerodynamic velocity V_0 (relative velocity of the airplane with respect to the air, assumed, for example, to be stationary at infinity) and the speed of sound at infinity upstream a_0.

The mechanics of flight immediately bring to mind the trajectory of aircraft and the catastrophic consequences that human errors or computer calculation errors can have on flight safety. The crash of the Indonesian Lion Air (October 2018) and Ethiopian Airlines (March 2019) Boeing 737 Max aircraft led to the deaths of 346 innocent people and a shutdown of around 800 such aircraft around the world.

A US Congress report highlights the heavy financial pressure on Boeing and the 737 Max program to get ahead of the release of Airbus' new aircraft, the A320 Neo, which affected the MCAS collision avoidance system implicated in the two accidents, as well as the culture of concealment that prevailed in the manufacturer and that prevented it from sharing crucial information with the authorities, its customers and the pilots of the 737 Max.

We must learn from the consequences of these disasters in order to prevent them from happening again, and engineers have a major role to play, in the sense that they are the guarantors of the safety and reliability of manufactured aircraft and must do everything possible to manage risks in order to minimize the occurrence of such disasters.

NOTE.– The units used in this book are those of the international metric system.

I.1. Definition of axes

Let us use the direct trihedron defined in Figure I.1.

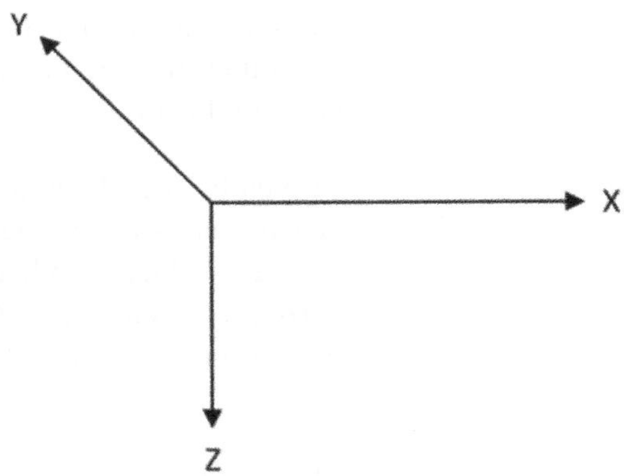

Figure I.1. *Direct trihedron*

Thus, the pitch, roll and yaw axes are defined as follows.

I.1.1. *Roll axis*

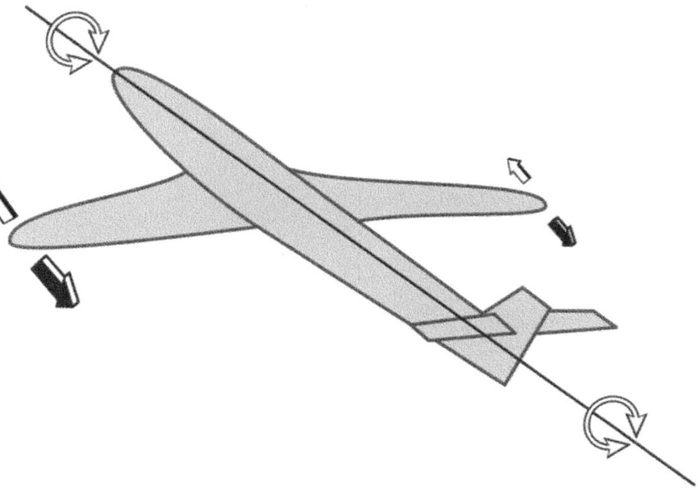

Figure I.2. *Roll axis (rotation around the X axis)*

I.1.2. *Yaw axis*

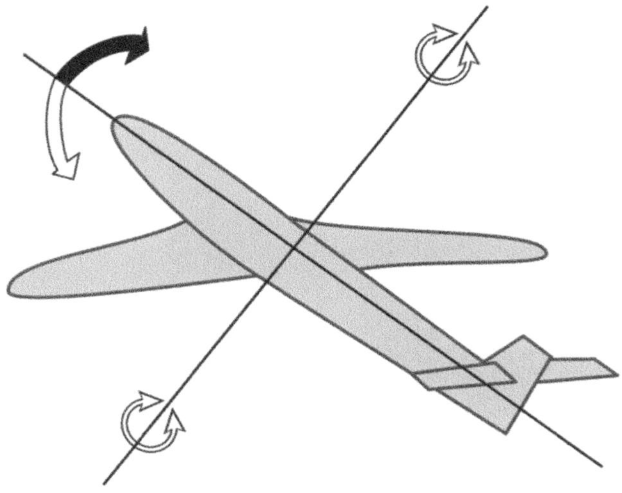

Figure I.3. *Yaw axis (rotation around the Z axis)*

I.1.3. *Pitch axis*

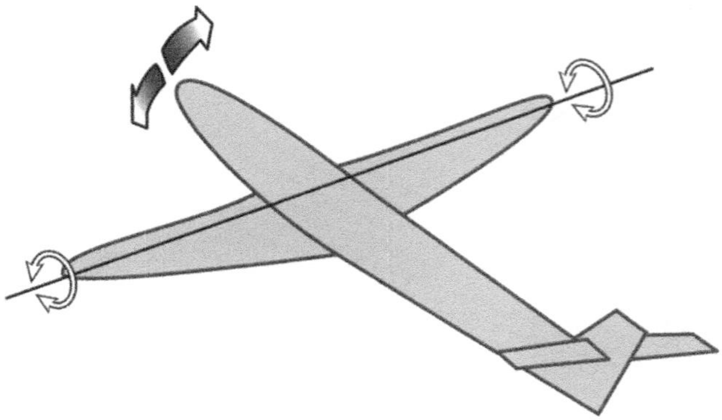

Figure I.4. *Pitch axis (rotation around the Y axis)*

1

Use of Aerodynamic Forces to Control the Trajectory of an Aircraft

In order to steer an aircraft in space, we use aerodynamic forces (lift and drag) created on small surfaces called control surfaces to produce rotations on the three axes of the aircraft (roll, pitch and yaw).

1.1. Definitions

Consider a thin, flat plate placed in a rapid air current (Figure 1.1).

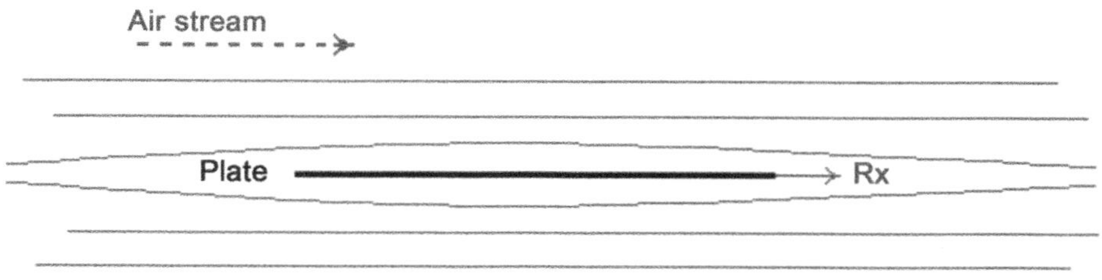

Figure 1.1. *Horizontal plate. For a color version of this figure, see www.iste.co.uk/louis/flight.zip*

In the diagram, the plate is horizontal: it does not tend to go up or down, but only moves back. If this plate were moving in still air instead of being held still in moving air, it would be slowed by a force called drag, conventionally denoted R_X. This force is due to the viscosity of the air, which sticks to the plate as it passes over and under it.

The plate is now tilted up. We gave it an angle with respect to the initial direction of the air streams (called the incidence α) (Figure 1.2).

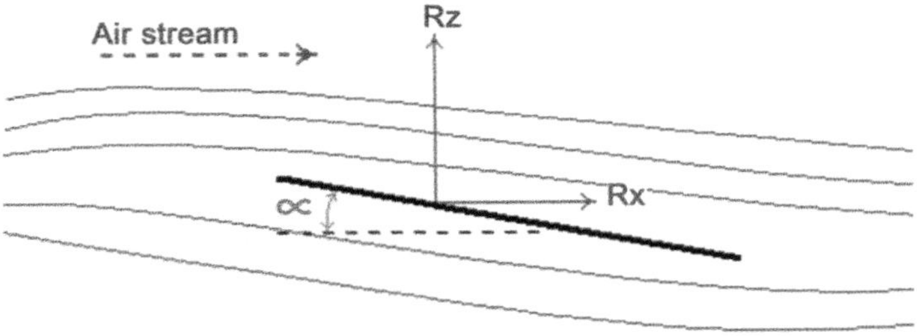

Figure 1.2. *Inclined plate. For a color version of this figure, see www.iste.co.uk/louis/flight.zip*

The air streams are deflected around the plate, which tends to rise and move back more sharply. The drag increases and a lift R_Z appears, pulling it upward. The combination of the two forces, lift and drag, combine in an aerodynamic resultant R.

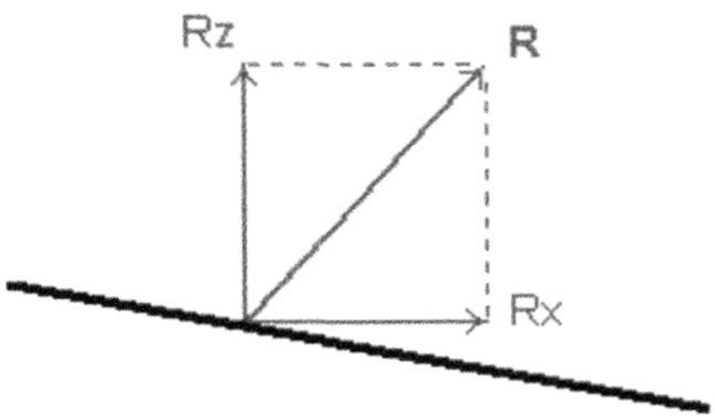

Figure 1.3. *Aerodynamic resultant. For a color version of this figure, see www.iste.co.uk/louis/flight.zip*

NOTE.– The indices x and z assigned to the aerodynamic forces come from a trihedron defined by convention around an aircraft in flight (Figure 1.4).

G being the center of gravity of the aircraft, there are three axes:

– the G_x axis, or roll axis;

– the G_y axis, or pitch axis;

– the G_z axis, or yaw axis.

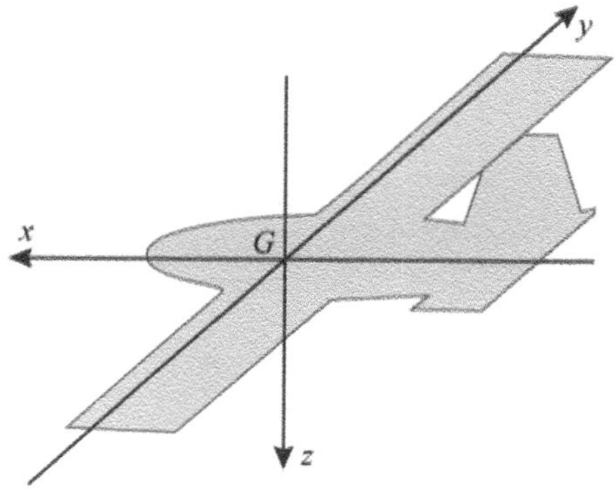

Figure 1.4. *Trihedron related to the aircraft's center of gravity*

As the aircraft turns on the G_x, G_y or G_z axis, it rolls, pitches or yaws, respectively.

If we consider a wing section (Figure 1.5) or profile, we can define some important terms for what follows.

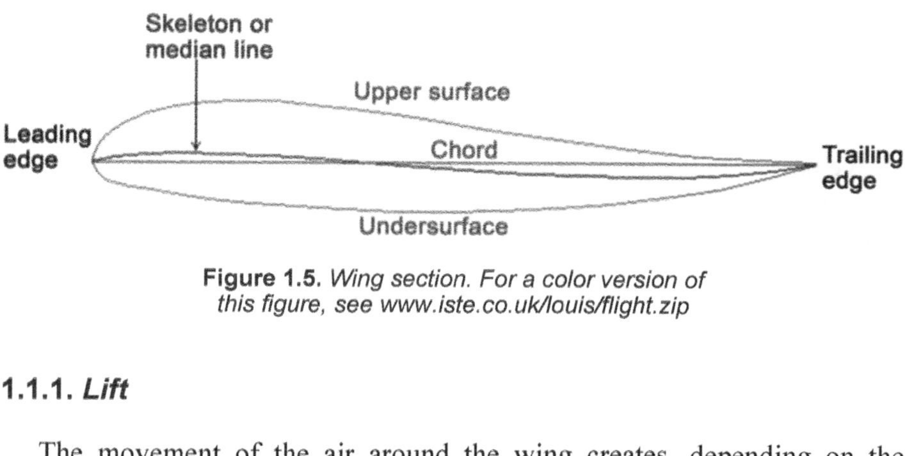

Figure 1.5. *Wing section. For a color version of this figure, see www.iste.co.uk/louis/flight.zip*

1.1.1. *Lift*

The movement of the air around the wing creates, depending on the profile and the angle of attack, a weak depression or overpressure under the wing, on the undersurface, and especially a strong depression above, on the upper surface.

It is the large pressure difference between the upper surface and undersurface that creates the lift.

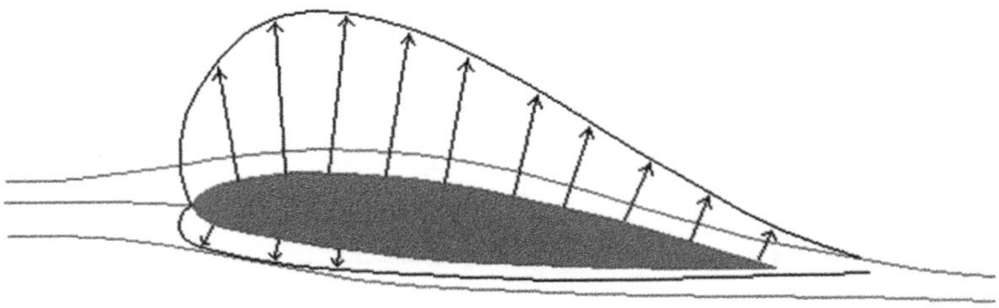

Figure 1.6. *Lift*

The wing moves forward with a certain angle of attack, which has the effect of deflecting the air downward. The air deflected violently downwards creates a void above, on the upper surface of the wing, which is thus sucked upwards. Air is a fluid that has a certain viscosity. This viscosity requires it to follow the profile; therefore, the air streams are separated by the wing, but they end up meeting after the trailing edge. When the wing passes, the air is therefore split in two: above and below.

The air below is deflected downwards, channeled as it is through the sloping undersurface. However, due to the viscosity of the air, the top streams also follow the profile. They stick to the upper surface, in exactly the same way a trickle of water follows the circular outline of a bottle placed under a faucet.

We have seen that lift is due to the intense downward deflection of the air and the viscosity of the air. Another phenomenon that contributes to the creation of dorsal wing depression is Bernoulli's principle, which says that accelerated air induces a drop in pressure (depression). This is what happens on the upper surface, where the air streams are accelerated as they try to travel a longer path (relative to the lower surface) in the same time.

1.1.2. *Drag*

The air flows along the surfaces and this friction produces significant braking. The airspeed is zero on the "skin" of the aircraft and it increases rapidly as soon as we move away from the surface. The layer of air close to the surface of the profile, where the air is slowed down, is called the boundary layer.

The shape of an object moving through air has a huge influence on induced drag. Anything that is smooth and tapered has little drag, because the air can move around the obstacle smoothly, effortlessly. Anything that disturbs the air and generates eddies is bad and increases drag.

The wing is a vortex generator. The pressure on the undersurface is much greater than on the upper surface. As a result, air passes from the lower surface to the upper surface at the ends and naturally to the trailing edge, to fill the void.

At the level of the wing tips, we can observe marginal vortices, which have a shape similar to that shown in Figure 1.7.

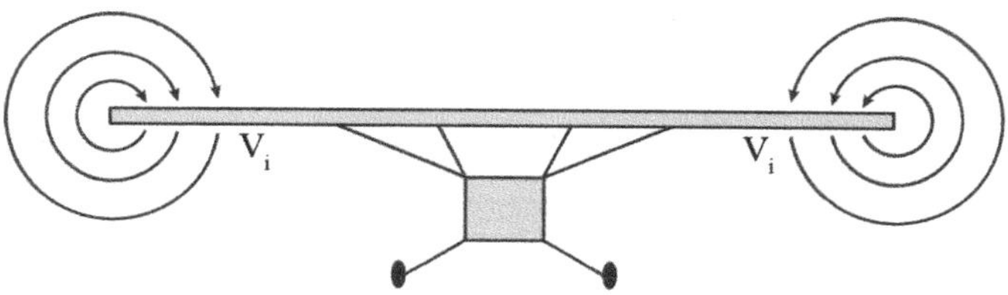

Figure 1.7. *Drag*

The passage of air from the bottom to the top of the wing creates a deflection of the air streams, outward of the wing for the lower surface and inward for the upper surface (Figure 1.8).

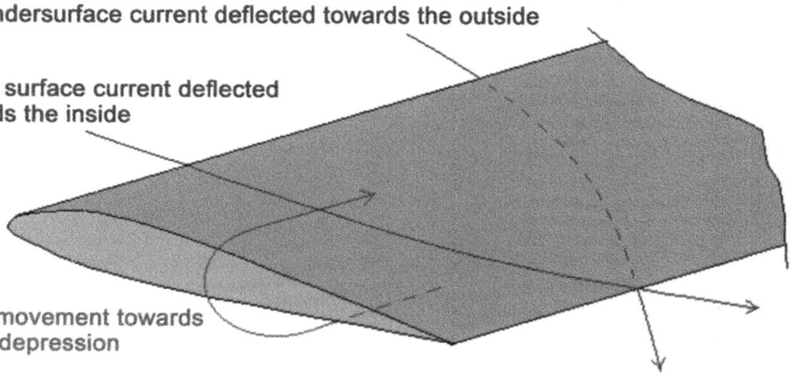

Figure 1.8. *Deflection of air streams. For a color version of this figure, see www.iste.co.uk/louis/flight.zip*

The intersecting air streams at the trailing edge also create vortices. To limit these marginal vortices, the wingtips are generally thinner. Thus, the passage of air from the bottom to the top is less abrupt. Some aircraft are equipped with flettners, kinds of vertical planes which reduce marginal losses and thus simulate a longer wing (the greater the aspect ratio, the more the influence of marginal losses decreases).

The marginal vortices increase drag and lower lift, creating airflow that rises from the lower surface to the upper surface and down to the wing. This descending air is animated by an induced velocity which is maximal at the wing tip and which decreases as we approach the center of the wing.

This induced velocity V_i has the effect of reducing the effective angle of attack of the wing towards the tips of the wings, which reduces lift.

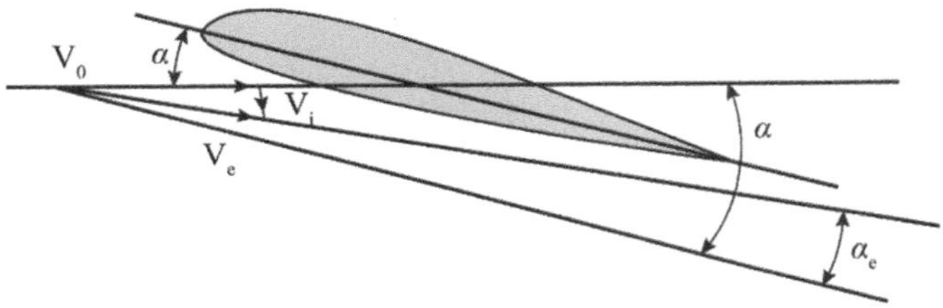

Figure 1.9. *Effects of induced velocity*

NOTE ON FIGURE 1.9.–

– V_0: *initial velocity*;

– V_e: *effective velocity*;

– V_i: *induced velocity*;

– α: *initial incidence*;

– α_e: *effective incidence*.

This ultimately results in a lift distribution over the wingspan in the shape of a semi-ellipse: the aircraft has a lot of lift in the center, and less and less the closer you get to the wing tips.

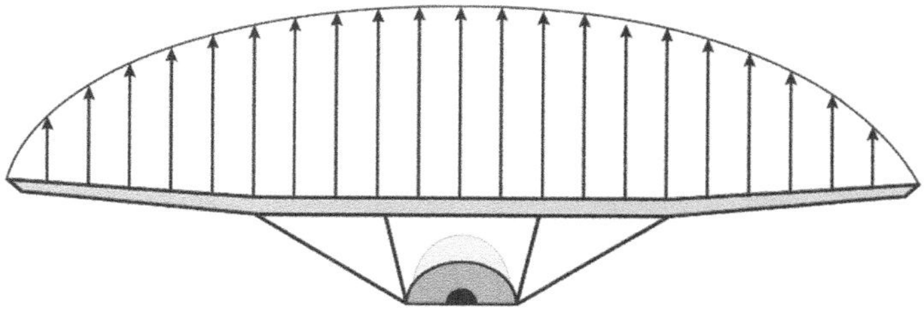

Figure 1.10. *Lift distribution*

1.1.3. *Equilibrium in horizontal flight*

Propeller pull opposes drag, just as lift opposes aircraft weight. In horizontal flight, we therefore have the equilibrium situation shown in Figure 1.11.

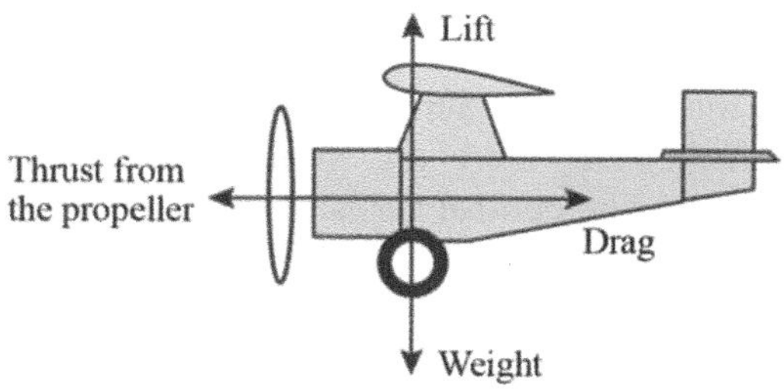

Figure 1.11. *Equilibrium in horizontal flight*

The lift is given only by the wing; on the other hand, the drag is induced by the wing itself as well as the fuselage, the antennas, the landing gear, etc. Ultimately, the wing is a small contributor to drag compared to everything else. Hence, we have the interest in the construction of aircraft of limiting as much as possible the relative importance of everything that does not produce lift. Making very long wings and a very thin, smooth fuselage increases the importance of lift over drag.

Here we come to the notion of the gliding ratio, which is the lift/drag ratio (or R_Z/R_X). It is also the maximum horizontal distance that a dropped aircraft can travel from a given altitude, divided by this same altitude: if the aircraft can travel a distance of 10,000 m when you stop the engine at an altitude of 1,000 m in calm weather, then it has a gliding ratio of 10,000 ÷ 1,000 = 10. This means that the lift of this aircraft is 10 times its drag.

1.1.4. *Aerodynamic moments*

A force F which is exerted at the level of the center of gravity of any solid body does not turn that body. If the point of application is moved away by a distance d, the aforementioned force produces an effect which will tend to set the body in rotation: it is said that it applies the moment F*d to the solid body in question.

In order to operate in all three dimensions, the aircraft experiences roll, pitch and yaw moments, which cause the latter to rotate around the axes defined in the introduction (section I.1) and which pass through its center of gravity.

We have seen that there are aerodynamic coefficients that apply to lift and drag forces. Likewise, there are moment coefficients that apply to aerodynamic moments. The control surfaces of an aircraft generate moments of roll (the ailerons of an aircraft), pitch (the front wing flaps or rear horizontal plane of an aircraft) or yaw (the rudder or rear horizontal plane of an aircraft). Wind tunnel testing quantifies the forces and moments applied to a wing profile or an entire aircraft. The most interesting aerodynamic moment is the pitch moment (whose moment coefficient is denoted C_m) because it is this which conditions the longitudinal balance.

1.1.5. *Center of gravity*

In physics, the center of gravity (CoG), denoted G, is the point of application of the resultant forces of gravity or weight. Therefore, it is clearly dependent on the gravitational field to which the body is subjected and should not be confused with the center of inertia, which is the barycenter of masses. It is often equated with the latter, but it is only an approximation linked to the fact that, in most cases, the gravitational field to which the body is subjected can be considered as uniform in the body considered.

In static, the center of gravity is the point of application of weight. This is a simplification of looking at the weight as a force applied to a single point, G, rather than considering a volume force applied to each point of the object.

In addition to simplifying the static calculations, knowledge of the position of the center of gravity is essential to determine the stability of an object:

– for an object placed on the ground, the action line of the weight must pass into the support polygon. If it is outside, the object tilts;

– when a vehicle accelerates or slows down, an object with a high center of gravity may tip over. This is more a property of the center of mass, and it follows from the principle of equivalence between gravity and acceleration;

– when lifting an object (slinging), the center of gravity is aligned with the vertical passing through the attachment point of the slings (straps or cables). If the point of attachment is not directly above the center of gravity at the start, the object swings;

– for a floating object, the center of gravity is positioned in line with the center of thrust of the pressure forces on the hull (Archimedean thrust). If the center of gravity shifts, the object tilts;

– when a single person lifts a load by hand, they must make sure that the center of gravity of the object is as close as possible to their pelvis and in particular must work with the back as straight as possible. This limits the flexion stress on the lumbar vertebrae.

1.1.6. *Aerodynamic center*

The aerodynamic center is defined as the point that does not cause a change in the pitch moment taken with respect to this point when the angle of attack varies. In other words, the aerodynamic center, which should not be confused with the center of thrust, is the point where the longitudinal moment is constant. This is also the point where the variations in lift (and not the lift itself) apply.

1.1.7. *Center of thrust*

By analogy, we can say that the center of thrust is to aerodynamic forces what the center of gravity is to gravitational forces. Indeed, when a solid is placed in a moving fluid, a force is exerted at every point of its surface (pressure or depression). The sum of all these forces is a single force, applied to the center of thrust of the solid.

The center of thrust of a wing profile is the point of application of lift. Its position varies (more or less depending on the type of profile) depending on the incidence.

1.1.8. *Centering and stability*

Whatever the profile adopted, it is the centering of the aircraft that conditions the longitudinal stability. Centering an aircraft means placing its center of gravity at a certain distance from the aerodynamic center. The absolute rule of stability is that the CoG should be in front of the aerodynamic center.

Here is a simplified chronology of the phenomenon of longitudinal stabilization:

1) the aircraft flies horizontally and in a straight line;

2) turbulence occurs; it increases the angle of attack of the aircraft;

3) as the angle of attack increases, the lift increases;

4) the increase in lift (ΔF_Z) occurs at the level of the aerodynamic center that is behind the CoG (because the aircraft is well centered);

5) this additional effort ΔF_Z creates a nose-down moment, which tends to decrease the incidence;

6) the process repeats.

The same process occurs vice versa for a gust decreasing the incidence.

It should be noted that there are aircraft designed to be naturally unstable: this is the case with modern fighter aircraft.

The Dassault Mirage 2000 and the Rafale, for example, were designed in this way, which makes them extremely maneuverable.

These aircraft are equipped with an artificial stabilization system using electric flight controls, controlled by an electronic computer. Without it, they would be very difficult or even impossible to pilot.

1.1.9. *Lateral stability*

The aircraft must have some roll stability. On an aircraft, the ailerons are used to restore the horizontal attitude.

However, it is good that the aircraft has, independently of any action by the pilot, a natural stability which results in the ability to automatically return to a position of equilibrium. This property is produced by the dihedral of the wing, i.e. the angle that the wings make with the horizontal, and possibly by the high position of the wing (CoG below the airfoil).

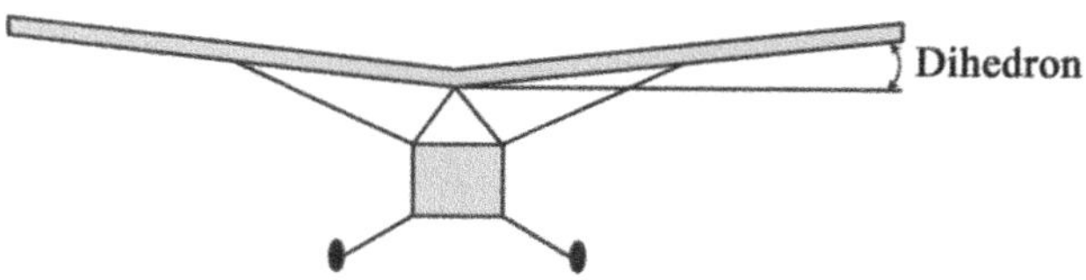

Figure 1.12. *Lateral stability*

The dihedral can be positive (wing facing up), zero (wing horizontal) or negative (wing pointing down). A positive dihedral is a source of roll stability, a zero dihedron is neutral and a negative dihedron is unstable (fighter aircraft, for which maximum maneuverability is required).

With a positive dihedral aircraft, if one wing lowers, it approaches a horizontal position and its lift combined with the weight of the aircraft creates more righting torque than the torque that is generated by the lift of the other wing. A light aircraft must have a more pronounced (and positive) dihedral the lower its wing. With a high wing, the CoG is below the airfoil and the whole behaves a bit like a parachute, naturally stable.

1.1.10. *Turn and roll*

With a conventional aircraft, especially below a certain speed, it is common to have to accompany the rudder with an action on the ailerons to put the aircraft in roll (when the speed increases, the action of the control stick on the ailerons may be sufficient).

However, the action on the rudder can be accompanied by induced roll:

– the aircraft initiates a flat change of direction;

– the outer wing in the turn sees its speed increase, so its lift also increases (see the formula for lift in section 2.2.3: it varies according to the square of the speed);

– simultaneously, the inner wing sees its speed decrease and therefore its lift as well;

– the aircraft gets into a roll.

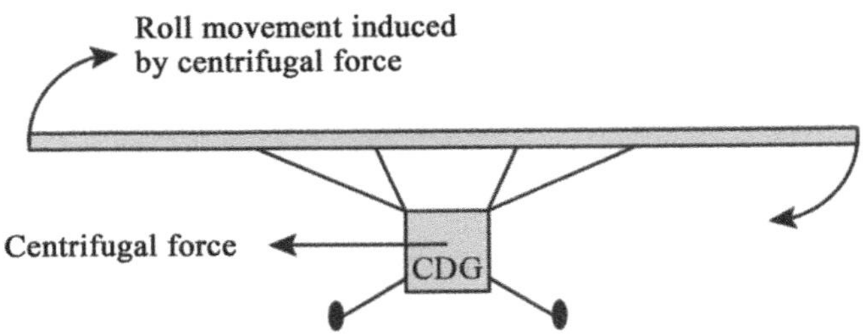

Figure 1.13. *Induced roll motion*

Hence, you have to pull the control stick if you want to tighten a turn.

1.1.11. *Load factor*

The load factor can be defined by the apparent weight of the aircraft divided by the actual weight. In a curved trajectory (turn, resource), the inertia of the aircraft makes it appear heavier (and the pilot also). If it appears twice as heavy, we speak of a load factor of 2, or we say that we are at 2 g, with g being the acceleration of the earth's gravity; in a turn where the aircraft tilts at an angle of X degrees, the load factor experienced is equal to $(1/\cos X)$.

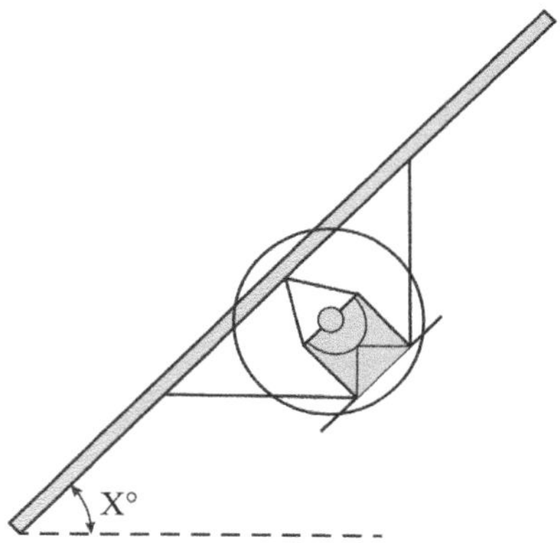

Figure 1.14. *Load factor*

EXAMPLES.–

– At 60° inclination, cos(60°) = 0.5, so the load factor is 2 g.

– At 80°, the load factor is almost 6 g.

1.2. Pitch control

Initially, for balanced level flight (Figure 1.16), the horizontal stabilizer of the aircraft is a spoiler (i.e. its lift is oriented downward). Changes to the lift of the horizontal stabilizer are achieved by turning the elevator. This makes it possible to modify the profile and the incidence of the horizontal stabilizer.

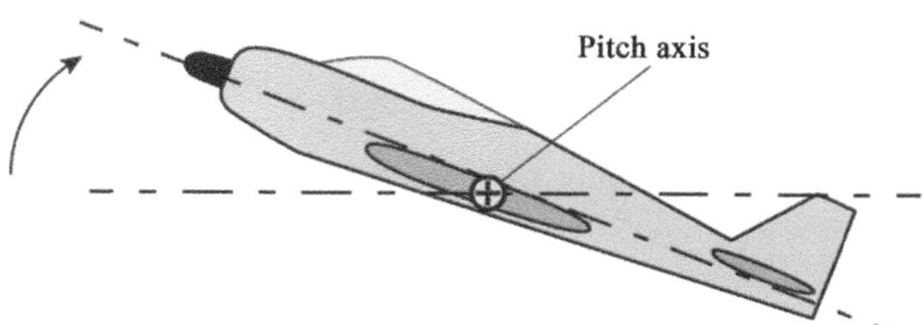

Figure 1.15. *Pitch control*

The variations in lift are then obtained simply by varying the incidence of the rudder. Elevator movements are symmetrical on the right and left sides of the horizontal stabilizer.

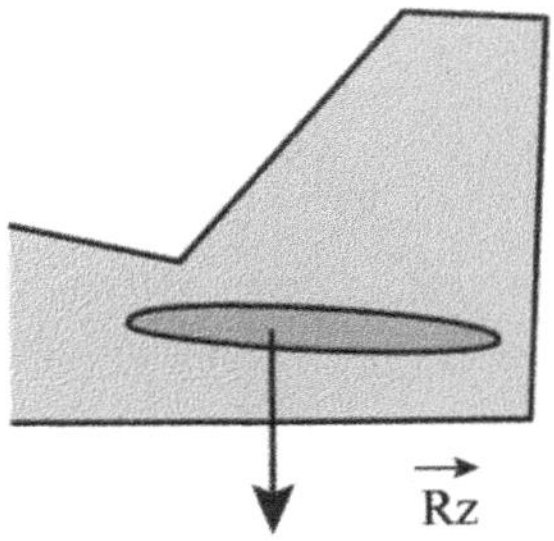

Figure 1.16. *Balanced level flight*

To move the aircraft up, you have to tilt its nose skyward, and to bring it down, you have to tip it towards the ground. In other words, it is necessary to cause a rotation around the pitch axis (see section I.1). In practice, we play on the lift of the horizontal stabilizer. If we increase its lift, the tail of the aircraft will rise and the nose will tilt towards the ground, so the aircraft will descend (Figure 1.17).

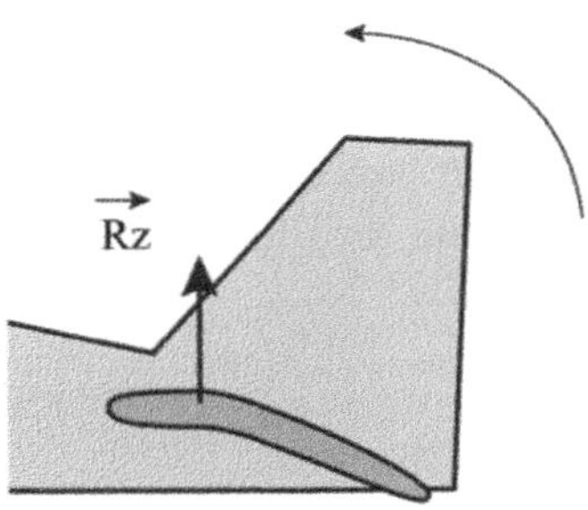

Figure 1.17. *Negative pitch*

If you decrease its lift, the tail of the aircraft will sink and its nose will swing skyward, so the aircraft will climb (Figure 1.18).

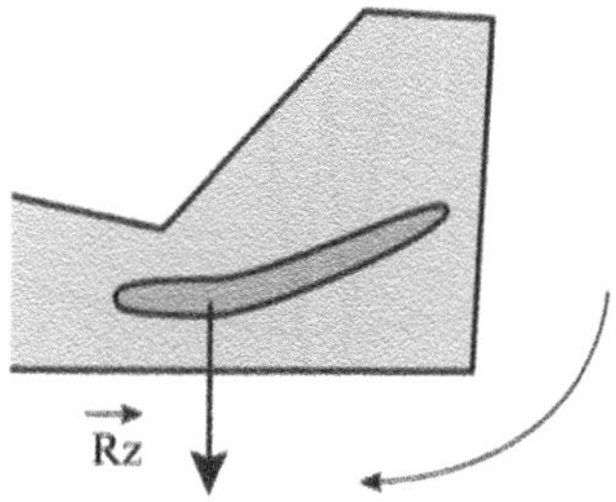

Figure 1.18. *Positive pitch*

1.3. Yaw control

To turn the nose of the aircraft to the left or to the right, it is necessary to cause a rotation around the yaw axis, i.e. the axis perpendicular to the plane of the wings and passing through the center of gravity of the aircraft (see section I.1). It is the vertical stabilizer that creates this rotation.

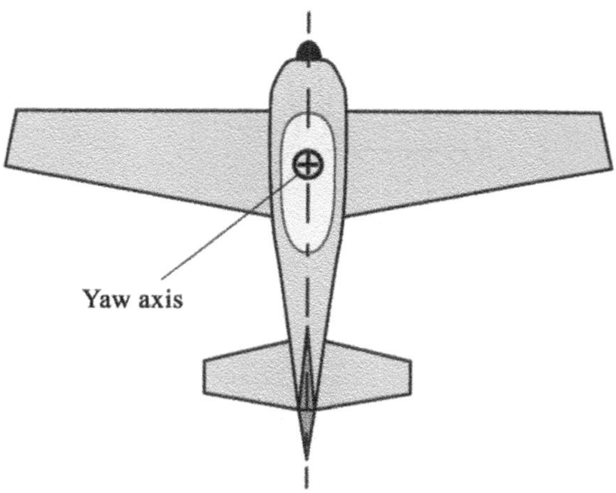

Figure 1.19. *Yaw control*

A vertical mobile surface, the rudder, creates an aerodynamic force, which causes the tail to rotate to the right or to the left. In the rest position, the rudder is in line with the aircraft.

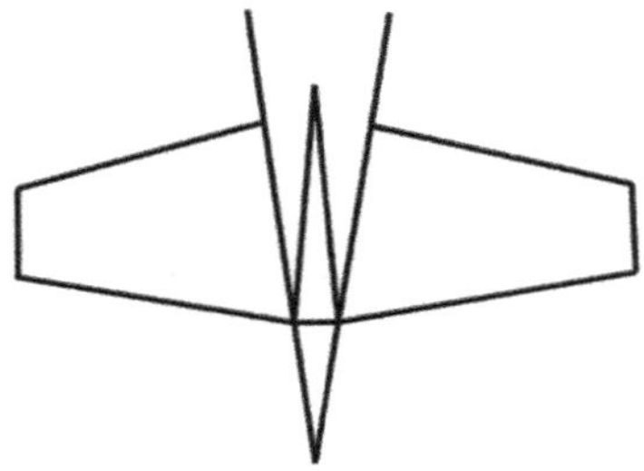

Figure 1.20. *Zero yaw*

The drift creates drag in the axis of the aircraft. When the rudder is turned to the right (by pressing the rudder to the right), an aerodynamic result is generated to the left on the vertical tail (Figure 1.21). It is therefore pulled in that direction and the nose of the aircraft goes to the right.

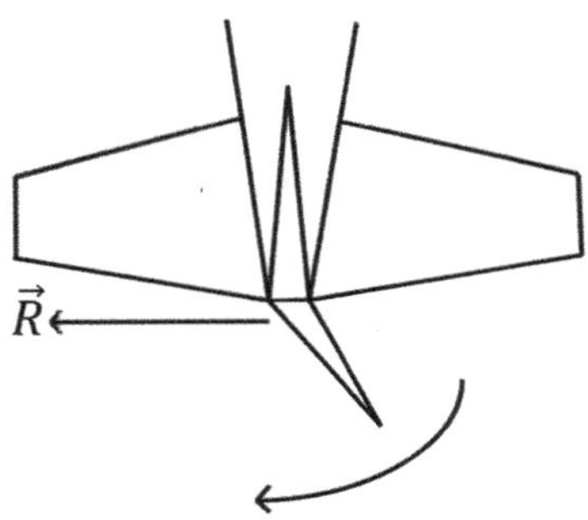

Figure 1.21. *Right yaw*

If we steer the rudder to the left, pushing the rudder to the left, we generate an aerodynamic result on the vertical tail unit directed to the right (Figure 1.22). The tail therefore goes to the right and the nose of the aircraft to the left.

In practice, to bring the nose to the right or to the left, you have to insert your foot on the side where you want to bring the nose.

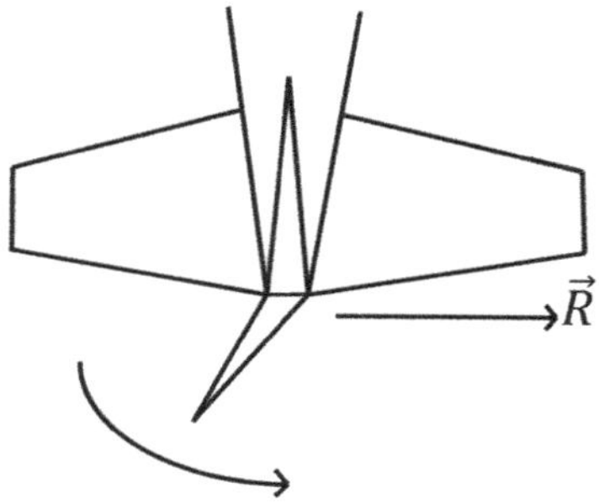

Figure 1.22. *Left yaw*

1.4. Roll control

To tip the aircraft onto the left wing or the right wing, you have to adjust the lift of each wing. In other words, to cause a rotation around the roll axis (see section I.1), which joins the nose of the aircraft to the tail, it is necessary to increase the lift of one wing and decrease that of the other.

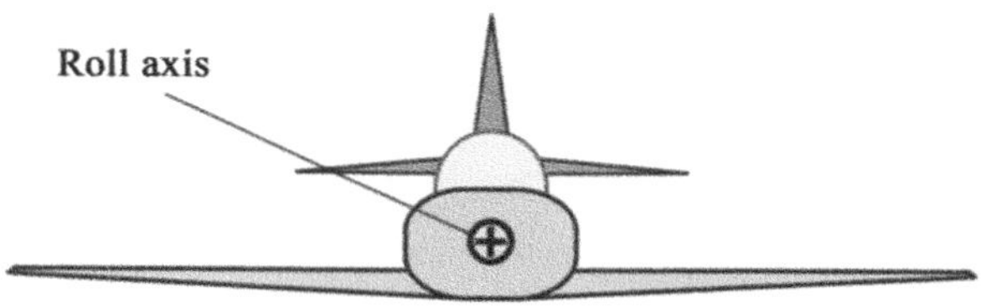

Figure 1.23. *Roll control*

To bank to the right, you need to decrease the lift on the right wing and increase the lift on the left wing. To bank to the left, you have to do the opposite. To change the lift of the wings, you use ailerons. These control surfaces, located at the ends of the wings to increase their efficiency, make it possible to obtain an opposite effect on the two wings with a differential deflection: the aileron is lowered on the side where the lift is to be increased and it is raised on the side where it must be reduced. Lowering an aileron makes the profile more arched at the end of the wing and also increases its incidence. The result is an increase in wing tip lift. The opposite turn on the other wing has the opposite effect.

In summary, to tilt the aircraft to the right, you have to raise the right aileron and lower the left aileron (with the stick to the right) (Figure 1.24).

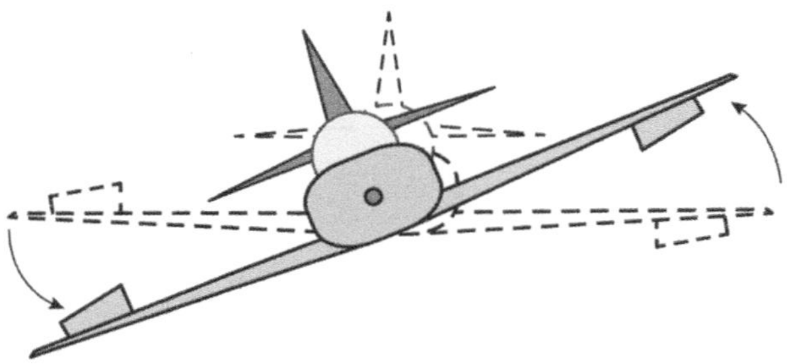

Figure 1.24. *Right roll*

To tilt the aircraft to the left, raise the left aileron and lower the right aileron (with the stick to the left (Figure 1.25).

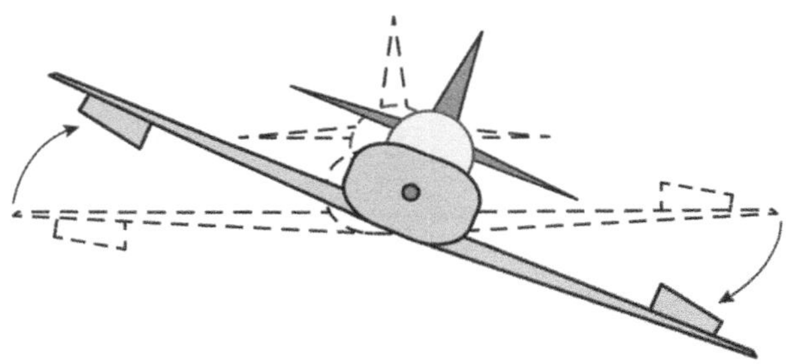

Figure 1.25. *Left roll*

1.5. The Mach: subsonic, transonic and supersonic speeds

It is said that an aircraft flies at subsonic speed when its speed is much lower than that of sound, i.e. less than 300 m/s. It is said that an aircraft flies at transonic speed when its speed is between 300 m/s and 380 m/s. Finally, an aircraft is said to fly at supersonic speed when its speed exceeds 380 m/s, which corresponds to a number greater than the higher critical Mach number.

The Mach number (M) is the ratio of the aerodynamic speed of the aircraft (V_{ae}) to the speed (C) of sound in air: $M = V_{ae}/C$ (see section 7.6). The speed of sound in air depends on the temperature:

– below M = 0.8, the speed is said to be subsonic;

– for $0.8 < M \leq 1.2$, it is transonic;

– for $1.2 < M < 5$, it is said to be supersonic;

– beyond that, it is hypersonic.

The transonic regime is difficult to model and control. It has a big impact on the aerodynamic behavior of the aircraft.

In this regime, the drag increases sharply as the lift drops, which strongly impacts the maneuverability and handling of the aircraft. Hence, this is a transitional regime but never a cruising regime.

2

Aerodynamic Forces and Moments and their Derivatives

The components of aerodynamic forces and moments are written as dimensionless coefficients called aerodynamic coefficients.

2.1. Definitions

The aerodynamic pressure is defined by the following equation:

$$P = \frac{1}{2} * \rho * V^2$$

with:

– P: the density of the air undisturbed by the aircraft;

– V: the relative aircraft speed with regard to the air speed, air undisturbed by the aircraft;

– S_{ref}: the reference surface, generally the surface of the maximum cross-section;

– L_{ref}: the reference length, generally the diameter of the body of the aircraft.

The aircraft's drag force (R_x) is positive when directed from the center of gravity to the rear of the aircraft.

The aircraft's lateral lift force (R_y) is positive when directed to the right.

The vertical lift force of the aircraft (R_z) is positive when it is directed along the descending vertical.

The torque (GM_x) is positive when it tends to give positive roll, i.e. this torque turns the aircraft to the right (if we look at the base of the aircraft).

The torque (GM_y) is positive when it tends to give nose-up pitch torque (positive nose-up with rising aircraft).

The torque (GM_z) is positive when it tends to turn the aircraft to the right in the yaw plane (positive yaw to the right of the aircraft).

2.2. Aerodynamic forces

2.2.1. *Drag expression*

$$R_x = \frac{1}{2} * \rho * V^2 * S_{ref} * C_x$$

The drag force is positive when it is directed from the center of gravity towards the rear of the aircraft; in the force balance, we associate it with the multiplying factor of (−1).

The drag of a profile results from the compressive forces in the axis of the aircraft. However, it can be broken down into three distinct parts:

– *form* drag;

– *wake* drag;

– *induced* drag (due to lift).

The first is related to the shape of the profile. The different profiles generate different flows. The pressure differences between the front and the rear are not identical, so the drag is influenced by the shape of the profile.

The second is related to the detachment of the air streams at the rear of the profile. The more the streams come off, the greater the wake drag. It is influenced by the speed and angle of flight of the aircraft.

The last is related to the pressure differences between the undersurface and the upper surfaces of the wing (which generate lift).

As the air from the undersurface is overpressured compared to that from the upper surface, the air from the underside of the airfoil tends to rise upwards at the wing tips. This creates eddies called marginal eddies.

At the end of the wing, they start from the undersurface of the wing and go up towards the upper surface. They widen as one gets further behind the wing.

The aircraft leaves behind two vortices that cause disturbances, which are all the more significant when the aircraft is large.

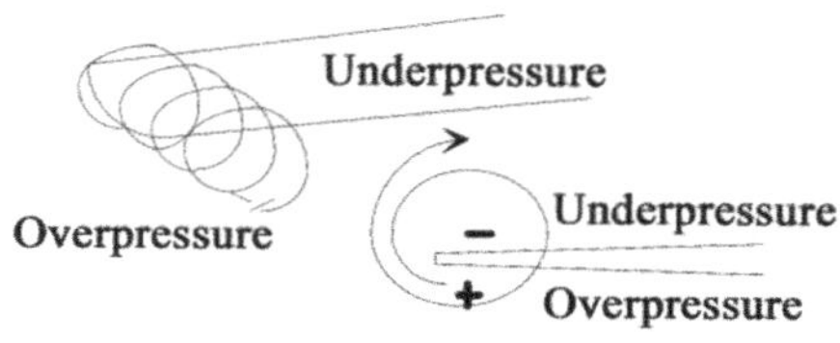

Figure 2.1. *Marginal vortices*

These can be very dangerous when passing in the wake of another aircraft, especially during take-off or landing.

This is the reason why, when two aircraft follow each other during take-off or when they approach an airport, they must be separated according to their respective weights (because the weight is representative of the size of the aircraft and therefore the power of the marginal vortices that it generates).

The same approach makes it possible, for a given profile, to study the influence of speed on drag. Measurements show that the higher the speed, the more the drag increases with the angle of attack.

Unlike lift, it increases continuously with the angle of attack.

The aspect ratio of an aircraft is the ratio between the square of its wingspan (L) and its wing area (S = wing area):

$$\lambda = \frac{L^2}{S}$$

Experience shows that the greater the aspect ratio, the lower the drag.

In fact, the aspect ratio influences the induced drag to a great extent. The quest for minimum drag and optimum lift sometimes requires high-aspect-ratio wings.

This is the case for airliners and gliders.

However, the high-aspect-ratio wings result in greater inertia and warping in the aircraft and therefore poorer agility.

C_x is the drag coefficient of the wing. It depends on the shape of the profile and the incidence of flight. Thin profiles have better C_x than thick profiles. C_x increases continuously with incidence, even after the incidence of stalling. Beyond this angle of attack, the drag continues to increase as the lift drops.

Drag is represented at the center of thrust, like lift.

2.2.2. *Expression of lateral lift*

The lateral lift force is given by the expression:

$$R_y = \frac{1}{2} * \rho * V^2 * S_{ref} * C_y$$

The lateral lift force is positive when it is directed to the right of the aircraft; in the force balance, we will associate with it the multiplying factor of (+1).

2.2.3. *Expression of lift*

The vertical lift force, more commonly called the lift force, is given by the expression:

$$R_z = \frac{1}{2} * \rho * V^2 * S_{ref} * C_z$$

The vertical lift force is positive when it is directed along the ascending vertical; in the balance of forces, we will associate with it the multiplying factor of (−1).

C_z is the wing's lift coefficient. It depends on the shape of the profile and the incidence of flight. Profiles with large curvatures have good C_z. C_z increases with the incidence up to the stall incidence. Once it exceeds this incidence, its value drops.

Lift is created on every part of the wing. To simplify its representation, it is materialized at a point by grouping together all of the small lift elements.

The point of application of overall lift is called the center of pressure (CP).

During the various phases of flight, the CP moves along the wing according to the speed and the angle of attack.

The greater its movements, the more difficult the aircraft will be to pilot.

The variations in lift are applied at another point called the focus (F). This is between a quarter (25%) and a third (33%) of the chord, starting from the leading edge. Its movements are limited.

2.3. Aerodynamic moments

The aerodynamic moments are given by the following three formulas:

$$GM_x = R_x * L_{ref} = \frac{1}{2} * \rho * V^2 * S_{ref} * L_{ref} * C_l$$

$$GM_y = R_y * L_{ref} = \frac{1}{2} * \rho * V^2 * S_{ref} * L_{ref} * C_m$$

$$GM_z = R_z * L_{ref} = \frac{1}{2} * \rho * V^2 * S_{ref} * L_{ref} * C_n$$

2.4. Major aerodynamic derivatives

The most commonly used aerodynamic derivatives are those relating to the angle of attack and sideslip, rudder deflection and the pitch, roll and yaw speeds.

The equations for the pitch coefficient are as follows:

– derivative with respect to the incidence:

$$C_{m\alpha} = \frac{\partial C_m}{\partial \alpha} \text{ and } C_{z\alpha} = \frac{\partial C_z}{\partial \alpha} \quad [2.1]$$

– derivative with respect to the pitch deflection:

$$C_{m\delta} = \frac{\partial C_m}{\partial \delta} \text{ and } C_{z\alpha} = \frac{\partial C_z}{\partial \delta} \quad [2.2]$$

– derivative with respect to the speed of rotation in pitch:

$$C_{mq} = \frac{\partial C_m}{\partial\left(\frac{q*L}{V}\right)} \text{ and } C_{z\delta} = \frac{\partial C_z}{\partial \delta} \quad [2.3]$$

The physical meanings of some of these terms are clear:

– the first represents the moment gradient with respect to the incidence, with a characteristic value for α around 0°;

– the second represents the efficiency of a pitch deflection;

– the third is roll damping.

3

Wind Tunnel

3.1. Description of wind tunnel tests

The lift of an aircraft depends on its speed. To understand this, just think of take-off. In order to obtain sufficient lift for take-off, the aircraft must reach a minimum lift speed.

We can thus say that, lift increases with speed. To verify this experimentally, we can place a model in a wind tunnel and use a balance to measure the lift (Figure 3.1).

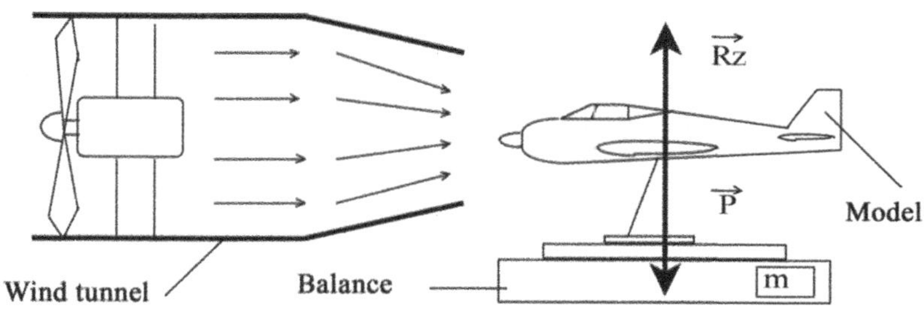

Figure 3.1. *Wind tunnel test*

The model and its support are placed in the test section. The support is based on a sensitive scale (at 0.01 g, for example) in order to record the variations in the apparent weight. When the wind tunnel is stopped, the scale is subjected to the weight of the model and the indicated mass m corresponds to that of the model. By taring the scale, the indicated mass changes to 0 g.

When the wind tunnel is started, a lift is created on the wings of the model and the scale indicates a negative mass. In fact, the scale is no longer subjected to the weight P of the model, but only to its apparent weight $P - R_z$. The mass indicated by the balance then translates the norm of the lift:

$$m = \frac{R_Z}{g}$$

The lift is given by the equation:

$$R_Z = m * g$$

Experience shows that m decreases (increasingly negative value) as the flow velocity increases. We can therefore deduce that the lift increases when the speed increases (Délery 2017).

3.2. Stall

The angle of attack is the angle formed by the direction of the airspeed and that of the aircraft heading line. This angle is of great importance as it greatly influences the lift of the wings of the aircraft.

The study of the evolution of lift as a function of the angle of attack is essential in aircraft design. To fly level at a given airspeed, the aircraft's lift must compensate for its weight. The speed being specified, the angle of attack of the flight is also fixed.

For an economical and comfortable flight, the angle of attack should be low for cruising speed.

We can see that, for a given speed, the lift increases with the angle of attack.

We can see that a fairly large negative impact generates negative lift. The sign of the mass measured by the scale has changes: the model presses on the scale with a force greater than its simple weight.

The low speeds of the wind tunnel and the small dimensions of the model only give qualitative results and make it difficult to demonstrate the stall.

We see in reality that, if the angle of attack exceeds a certain value, the lift no longer increases but drops very sharply. This is the wing stall (Figure 3.2).

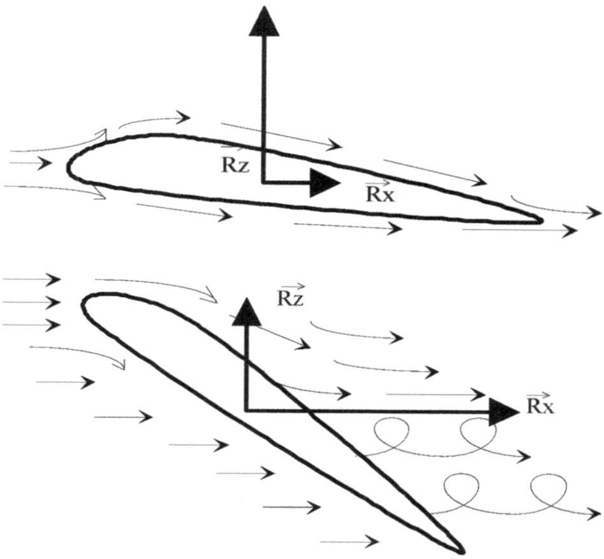

Figure 3.2. *Evolution of lift as a function of the angle of attack*

When the air is deflected by the profile, it adheres to the wing skin thanks to its viscosity. It therefore allows it to follow the curves of the undersurface and upper surface, thus generating lift.

If the angle of attack becomes too great, the air can no longer follow the curvature of the upper surface. The air streams come loose from the wing skin. The flow behind it is very turbulent, and pressure builds up on this part of the upper surface.

The lift decreases very sharply (the drag is significant) and the aircraft stalls. In conclusion, at a given speed, the lift of the wing increases with the angle of attack, up to the angle of stall.

The stall always occurs at the same angle of attack.

For the pilots of high-performance aircraft, the angle of attack is essential during the take-off and especially the landing phases.

In effect, fast aircraft are programed to cruise with low incidence (a few degrees at most). In the phases of flight at slow speed, in order to obtain sufficient lift, they need to increase their angle of attack (the use of high-lift devices makes it possible to limit this angle of attack but is not sufficient).

The drag is then significant and a sudden increase in the angle of attack can cause a stall, so it is necessary to be very careful.

Generally speaking, to maintain level flight, when the pilot increases speed, they have to decrease the angle of attack to maintain lift (they must therefore push lightly on the control stick). On the other hand, if they decrease the speed, they have to increase the angle of attack to maintain lift (they must therefore pull the stick slightly) (General Directorate of Civil Aviation 2012).

3.3. Calculation of the maximum stress applied to an aircraft wing in a wind tunnel

3.3.1. *Definition of the drag coefficient from the flow tables*

$$C_x = \frac{P_{i1} - P_0}{Q} = \frac{\frac{P_{i1}}{P_0} - 1}{\frac{Q}{P_0}} = \frac{\frac{P_{i1}}{P_0} - 1}{\frac{Q * P_{i0}}{P_{i1} * P_0}}$$

where:

– P_{i0} is the total or stagnation pressure before shock;

– P_{i1} is the total or stagnation pressure after shock;

– P_0 is the static pressure after shock;

– Q is the dynamic pressure before shock.

The following example is taken from Luneau 1974.

EXAMPLE.– Calculation of C_x at Mach 1.8:

$(P_{i1} / P_0) = 4.6695$

$(Q / P_{i0}) = 0.39472$

$(P_{i0} / P_0) = 0.17404$

$\rightarrow C_x = (4.6695 - 1) / (0.39472 / 0.17404) = 1.6180$

This coefficient is multiplied by the ratio of the surfaces S/L_{ref} (e.g. for the wings S_{bea}/L_{ref}).

3.3.2. *Calculation of the maximum stress applied to an aircraft wing in a wind tunnel*

The effort applied to a wing in a wind tunnel is given by the following equation:

$$F = \frac{1}{2} * \rho * V^2 * S_{ref} * C_{z(wings)}$$

where:

$$\frac{1}{2} * \rho * V^2 = \frac{\frac{1}{12}}{2} * \gamma * M^2 * P_{static}$$

Let us take the conditions of the model in the wind tunnel presented Figure 3.3.

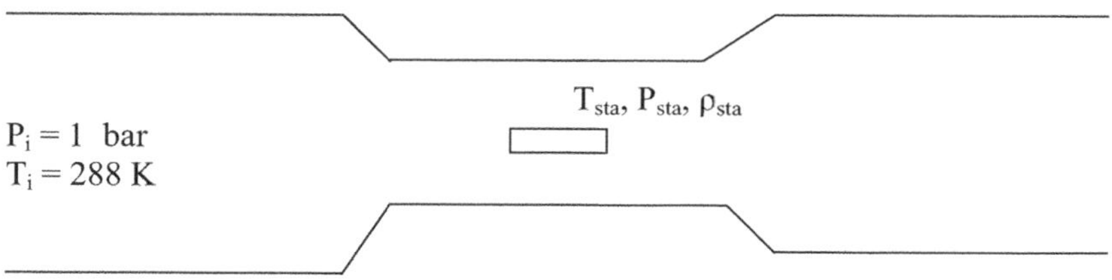

Figure 3.3. *Model in wind tunnel*

For M = 0.9:

– P_0: 0.59126 bar (static pressure);

– ρ: 0.68704 * 1.225 = 0.8416 kg/m³;

– T: 0.86058 * 288 = 247.8 K° (static temperature).

From the tables (see section 3.3.1):

$$\frac{Q}{P_i} = 0.33524 \text{ at Mach } 0.9 \; Q = 33{,}254 \text{ Pa}$$

for a C_{zmax} of the wings of 2.6/2 or 1.3.

The effort on the wing is given by:

$$F = \frac{1}{2} * \rho * V^2 * S_{ref} * C_{z(wings)} = 33{,}524 * 1.3 * S_{ref}$$

for S_{ref} = 0.0048707 F = 212.27 N.

4

Trihedron

4.1. Writing convention

The change of reference matrix making it possible to go from reference A to reference B in the axes of reference A is noted Matrix A/B in the rest of the book.

For example, the position of the aircraft in the aircraft frame E with respect to the local geographic frame can be written in the axes of the aircraft frame using the change of frame matrix Matrix $_{E/LGT}$ given in section 4.5.1.

4.2. Definitions of trihedrons

4.2.1. *Definition of a trihedron or frame of reference*

The association of the space frame $(O, \vec{i}, \vec{j}, \vec{k})$ and the time frame constitutes a frame of reference (R).

In the remainder of the book, we will note, for simplicity, the frame of reference (R) as follows:

$$(O, \overrightarrow{i_R}, \overrightarrow{j_R}, \overrightarrow{k_R})_R$$

4.2.2. *Galilean trihedron*

In physics, a Galilean or inertial frame of reference is a frame in which an isolated object (on which no force is exerted or on which the resultant of the forces is zero) is in uniform rectilinear translational movement (immobility being a particular case of uniform rectilinear movement): the speed of the body is constant (over time) in direction and in norm.

This means that the principle of inertia, which is stated in Newton's first law – "every body perseveres in the state of rest or of uniform straight-line motion in which it finds itself, unless some force acts upon it and compels it to change its state" – is true.

A Galilean referential is thus called as such as a tribute to Galileo and more particularly to Galilean relativity.

The search for an inertial frame of reference is a delicate subject; the concrete determination of such a frame of reference is always approximate.

In a non-inertial frame of reference, which is animated by an accelerated movement relative to a Galilean frame of reference, inertial forces must intervene.

4.2.3. *Absolute trihedron*

We assume that there are at least three stars, E1, E2 and E3, considered as points, which, together with the center of mass S of the solar system, form an undeformable tetrahedron.

Let $(O,\vec{i},\vec{j},\vec{k})$ be an orthonormal coordinate system rigidly linked to this tetrahedron. By providing this frame of reference with a time base, we constitute the Kepler frame of reference (K), which we assume to be Galilean.

This trihedron is also called the Copernicus or Kepler trihedron.

4.2.4. *Local geographic trihedron*

The Earth being assimilated to a sphere with center G, the frame of reference (LGT) is constituted using the frame of reference (OX_T, OY_T, OZ_T) defined as follows:

– O is a point on the surface of the Earth;

– OX_T is tangent to the meridian and directed towards the north;

– OY_T is tangent to the parallel of O and directed towards the east;

– OZ_T is directed towards the center of the Earth.

4.2.5. *Terrestrial trihedron*

The Earth being assimilated to a sphere of center G, the terrestrial reference frame (TER) is defined as follows:

– G is the center of the Earth;

– G_{X_T} is tangent to the meridian and directed towards the north;

– G_{Y_T} is tangent to the parallel of G and directed towards the east;

– G_{Z_T} completes this direct trihedron (directed towards the center of the Earth).

4.2.6. *Aircraft trihedron*

The aircraft trihedron (drone, helicopter, plane, etc.), or Lilienthal axis system (G, X_e, Y_e, Z_e), is defined as follows:

– the trihedron has an origin G, which is the center of gravity of the aircraft;

– GX_e is the longitudinal axis, which is carried by the axis of the aircraft towards the nose of the aircraft;

– GY_e is the axis perpendicular to the plane of symmetry, oriented to the right (pilot direction);

– $GZ_e \perp GX_e$ is located in the plane of symmetry of the aircraft and oriented > 0 towards the belly of the aircraft (towards the center of the Earth); it completes the direct trihedron.

4.2.7. *Aircraft aerodynamic speed trihedron (G, X_{vae}, Y_{vae}, Z_{vae})*

The aerodynamic speed trihedron of the aircraft $(G, X_{vae}, Y_{vae}, Z_{vae})$ is defined as follows:

– the trihedron has an origin G, which is the center of gravity of the aircraft;

– GX_{vae} is carried by the aerodynamic speed;

– $GZ_{vae} \perp GX_{vae}$ is located in the plane of symmetry of the aircraft and oriented > 0 towards the belly of the aircraft (towards the center of the Earth);

– GY_{vae} completes the direct trihedron.

4.2.8. *Balance trihedron*

The coefficients on tracks y (GY_b) and z (GZ_b) are restored in an aircraft trihedron demodulated of the aerodynamic roll. The balance trihedron is defined as follows (G, X_b, Y_b, Z_b):

– the trihedron has an origin G, which is the center of measurement of the balance;

– GX_b is carried by the X axis of the aircraft;

– GZ_b is located in the plane of symmetry of the aircraft and oriented > 0 towards the belly of the aircraft (towards the center of the Earth);

– GY_b completes the direct trihedron.

4.3. Change of reference

The trihedrons presented in Figures 4.1–4.3 make it possible to visualize the rotations applied to the machine, as well as the projections on the various planes of symmetry of the aircraft.

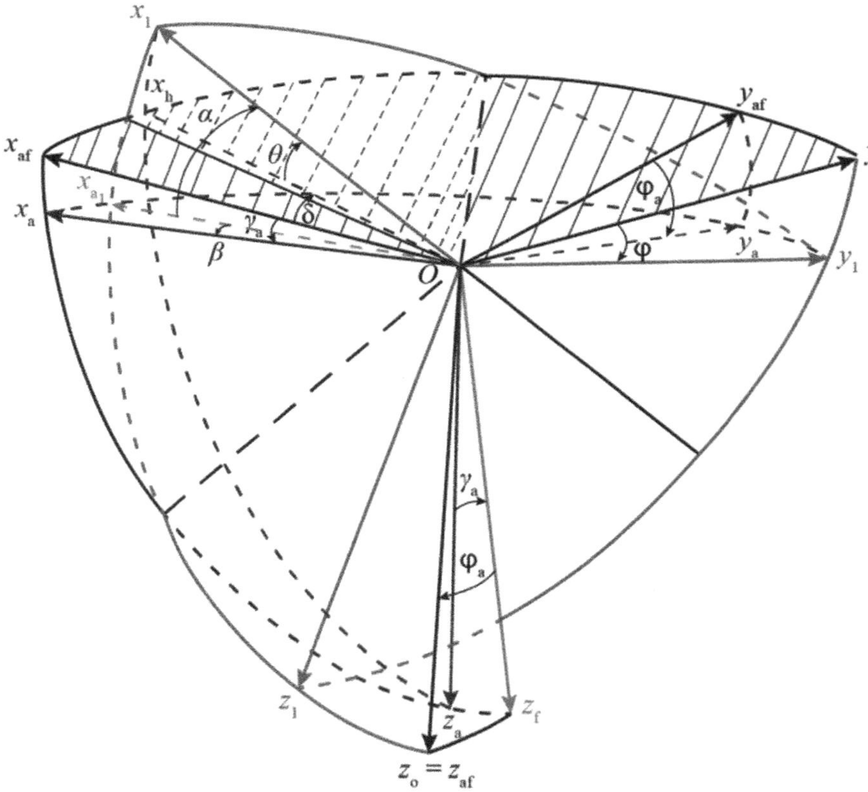

Figure 4.1. *Reference trihedra of the machine. Passage from the aircraft reference to the local geographical reference. $\beta < 0$; $\gamma_a < 0$. For a color version of this figure, see www.iste.co.uk/louis/flight.zip*

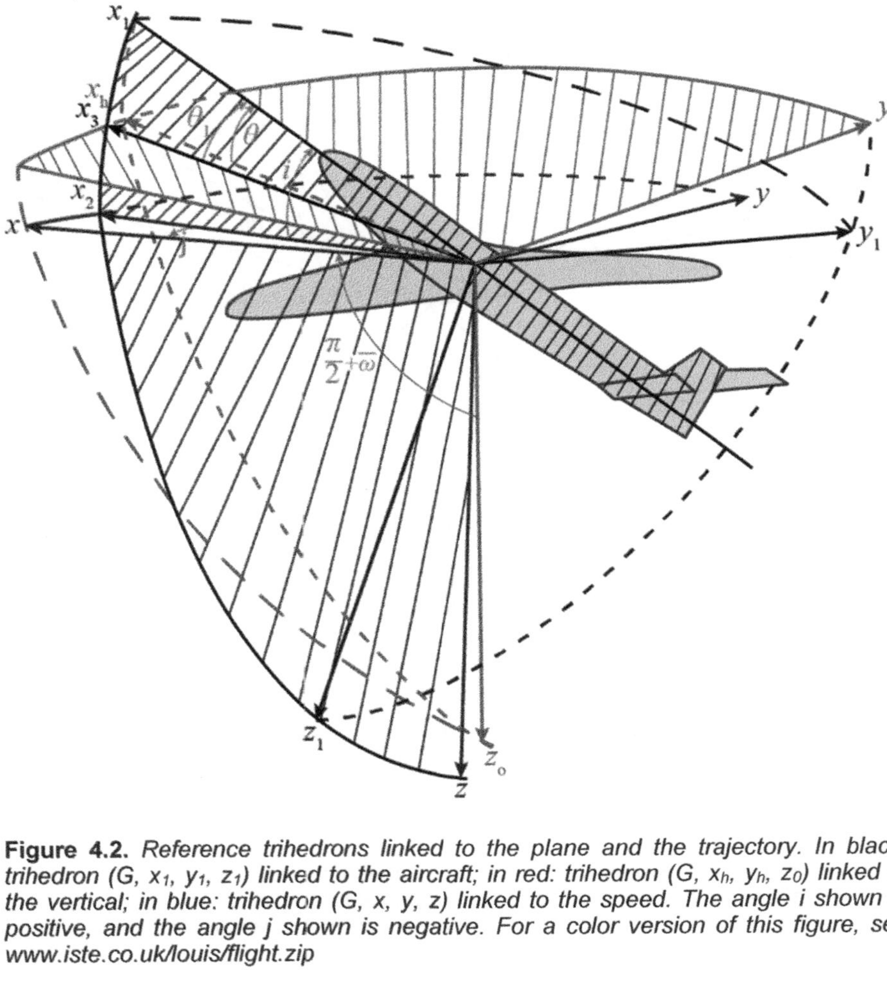

Figure 4.2. *Reference trihedrons linked to the plane and the trajectory. In black: trihedron (G, x_1, y_1, z_1) linked to the aircraft; in red: trihedron (G, x_h, y_h, z_0) linked to the vertical; in blue: trihedron (G, x, y, z) linked to the speed. The angle i shown is positive, and the angle j shown is negative. For a color version of this figure, see www.iste.co.uk/louis/flight.zip*

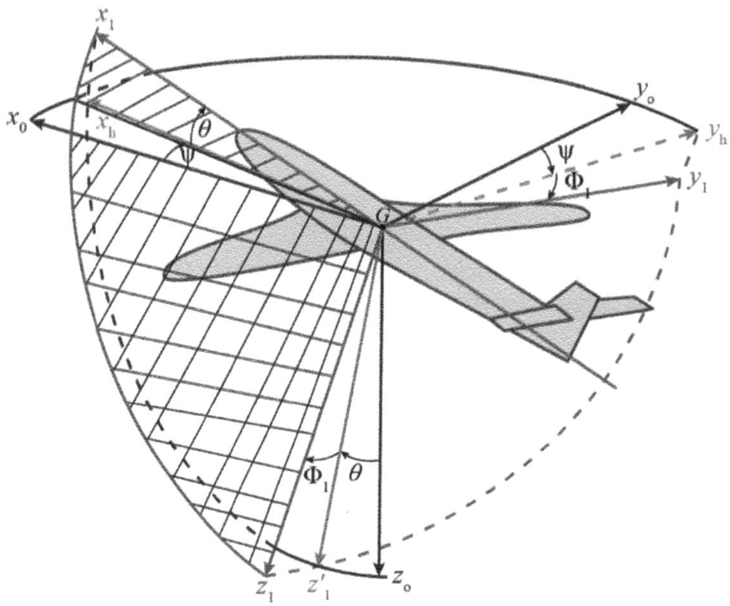

Figure 4.3. *Reference trihedra defining the angular position of the aircraft in space $\Phi_1 = \Phi$. In blue: the reference trihedron (G, x_0, y_0, z_0), Gz_0 depends on the gravity; in red: the plane trihedron (G, x_1, y_1, z_1). The hatched area is in the plane of symmetry of the aircraft. For a color version of this figure, see www.iste.co.uk/louis/flight.zip*

4.4. Relation between trihedra

4.4.1. *Aircraft trihedron (G, X_1, Y_1, Z_1) with respect to the reference trihedron (G, X_0, Y_0, Z_0)*

This position is defined by the coordinates of the center of gravity G and the relative orientation of the axes. From the coordinates of G, only the altitude Z is kept. The relative position is defined by Figure 4.3. We used, to define the angles, the trihedron (G, X_h, Y_h, Z_0), where GX_h is the projection of GX_1 on the horizontal plane passing through G:

– Ψ: the angle defining the azimuth (or the heading if OX_0 is directed towards the north);

– θ: the longitudinal attitude angle (also called the attitude);

– ϕ: the roll angle.

We still use the angles θ_1 and ϕ_2. (G, X, Z) is the intersection of the plane of symmetry(G, X_1, Z_1) with the horizontal plane of G:

– θ_1: the pitch angle;

– ϕ_2: the angle of lateral attitude (angle of $GY_1$with the horizontal plane not shown).

4.4.2. *Aerodynamic trihedron of the aircraft (G, X_{va}, Y_{va}, Z_{va}) with respect to the aircraft trihedron (G, X_e, Y_e, Z_e)*

The aerodynamic slope is defined by the sideslip angle β and angle of attack α:

– β: the angle of the speed vector with the plane of symmetry;

– α: the angle of the projection of the speed vector on the plane of symmetry with the longitudinal reference of the aircraft.

We define the slope as the angle of the velocity with the horizontal plane, counted positively when GX is located above the horizontal plane (see Figures 4.1 and 4.2; in Figure 4.2, the angle α is denoted as i and the angle β is denoted as j).

4.5. Relative positions of the various trihedra

4.5.1. *Position of the aircraft trihedron with respect to the local geographical trihedron*

The reference trihedron is the local geographic trihedron defined at the point of projection of the aircraft on the launch ramp $(G, X_{lgt}, Y_{lgt}, Z_{lgt})$.

For the change of trihedron, we carry out the three classic rotations:

– of the angle Ψ around the axis Z_{lgt}, which brings GX_{lgt} into GX_1 and GY_{lgt} into GY_1, in the vertical plane passing by GZ_{lgt};

– of the angle θ around the axis Y1, which brings GY_1 into GX_e and GZ_{lgt} into GZ_1;

– of the angle ϕ around the axis GX_e, brings GY_1 into GY_e and GZ_1 into GZ_e.

Consequently, the coordinates $(X_{lgt}, Y_{lgt}, Z_{lgt})$ in the local geographic trihedron are written using the following transfer matrix from the coordinates (X_e, Y_e, Z_e) in the aircraft trihedron:

$$\begin{pmatrix} 1 & 0 & 0 \\ 0 & 1 & 0 \\ 0 & 0 & 1 \end{pmatrix}$$

$$\begin{pmatrix} X_{lgt} \\ Y_{lgt} \\ Z_{lgt} \end{pmatrix} = \begin{pmatrix} \cos\psi & -\sin\psi & 0 \\ \sin\psi & \cos\psi & 0 \\ 0 & 0 & 1 \end{pmatrix} \begin{pmatrix} X_1 \\ Y_1 \\ Z_{lgt} \end{pmatrix}$$

$$\begin{pmatrix} X_1 \\ Y_1 \\ Z_{lgt} \end{pmatrix} = \begin{pmatrix} \cos\theta & 0 & \sin\theta \\ 0 & 1 & 0 \\ -\sin\theta & 0 & \cos\theta \end{pmatrix} \begin{pmatrix} X_e \\ Y_1 \\ Z_1 \end{pmatrix}$$

$$\begin{pmatrix} X_e \\ Y_1 \\ Z_1 \end{pmatrix} = \begin{pmatrix} 1 & 0 & 0 \\ 0 & \cos\phi & -\sin\phi \\ 0 & \sin\phi & \cos\phi \end{pmatrix} \begin{pmatrix} X_e \\ Y_e \\ Z_e \end{pmatrix}$$

$$\begin{pmatrix} X_{lgt} \\ Y_{lgt} \\ Z_{lgt} \end{pmatrix} = \begin{pmatrix} \cos\psi * \cos\theta & -\sin\psi & \cos\psi * \sin\theta \\ \sin\psi * \cos\theta & \cos\psi & \sin\psi * \sin\theta \\ -\sin\theta & 0 & \cos\theta \end{pmatrix} * \begin{pmatrix} 1 & 0 & 0 \\ 0 & \cos\phi & -\sin\phi \\ 0 & \sin\phi & \cos\phi \end{pmatrix} \begin{pmatrix} X_e \\ Y_e \\ Z_e \end{pmatrix}$$

$$\begin{pmatrix} X_{lgt} \\ Y_{lgt} \\ Z_{lgt} \end{pmatrix} = (\text{Matrix}) \begin{pmatrix} X_e \\ Y_e \\ Z_e \end{pmatrix}$$

and:

$$\begin{pmatrix} X_e \\ Y_e \\ Z_e \end{pmatrix} = (\text{Matrix})^{-1} \begin{pmatrix} X_{lgt} \\ Y_{lgt} \\ Z_{lgt} \end{pmatrix}$$

with:

$$(\text{Matrix})_{\frac{\text{LGT}}{\text{Machine}}} = \begin{pmatrix} M_{11} & M_{12} & M_{13} \\ M_{21} & M_{22} & M_{23} \\ M_{31} & M_{32} & M_{33} \end{pmatrix}_{\frac{\text{LGT}}{\text{Machine}}}$$

The inverse matrix, which makes it possible to pass from the aircraft trihedron to the local geographical trihedron, is symmetrical compared to the diagonal:

$$(\text{Matrix}^{-1})_{\text{Machine/LGT}} = \begin{pmatrix} M_{11} & M_{21} & M_{31} \\ M_{12} & M_{22} & M_{32} \\ M_{13} & M_{23} & M_{33} \end{pmatrix}_{\text{Machine/LGT}}$$

where:

$$M_{11} = \cos\psi * \cos\theta$$

$$M_{12} = -\sin\psi * \cos\phi + \cos\psi * \sin\theta * \sin\phi$$

$$M_{13} = \sin\psi * \sin\phi + \cos\psi * \sin\theta * \cos\phi$$

$$M_{21} = \sin\psi * \cos\theta$$

$$M_{22} = \cos\psi * \cos\phi + \sin\psi * \sin\theta * \sin\phi$$

$$M_{23} = -\cos\psi * \sin\phi + \sin\psi * \sin\theta * \cos\phi$$

$$M_{31} = -\sin\theta$$

$$M_{32} = \cos\theta * \sin\phi$$

$$M_{33} = \cos\theta * \cos\phi$$

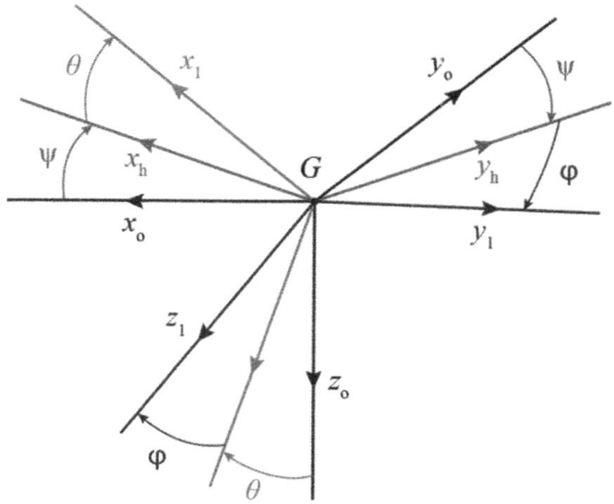

Figure 4.4. *Position of the aircraft trihedron with respect to the reference trihedron. The planes (G, x_O, y_O) and (G, x_h, y_h) are parallel. In black: the reference trihedron (G, x_O, y_O, z_O) with Gz_O directed towards the center of the Earth (axis of gravity). The machine trihedron is represented by (G, x_1, y_1, z_1). For a color version of this figure, see www.iste.co.uk/louis/flight.zip*

4.5.2. *Position of the aerodynamic trihedron with respect to the terrestrial trihedron*

The reference trihedron is the earth trihedron redefined with respect to the center of the Earth (G, X_0, Y_0, Z_0).

The aerodynamic trihedron (G, X_A, Y_A, Z_A) is usually identified by three classical angles compared to the reference trihedron.

For the change of trihedron, we carry out the three classic rotations:

– the aerodynamic route or the aerodynamic azimuth χ_a: rotation around axis Z_0, which brings GX_0 into GX_h and GY_0 into GY_h, in the vertical plane passing from GZ_0;

– the aerodynamic slope γ_a around the axis GY_h, which brings GX_h into GX_a and GZ_0 into GZ_f;

– the rotation μ_a around the axis GX_a, which brings GY_h into GY_a and GZ_f into GZ_a.

Therefore, the coordinates (X_A, Y_A, Z_A) in the terrestrial trihedron are written using the following transfer matrix from the coordinates (X_0, Y_0, Z_0) in the terrestrial trihedron:

$$\begin{pmatrix} X_0 \\ Y_0 \\ Z_0 \end{pmatrix} = \begin{pmatrix} \cos\chi_a & -\sin\chi_a & 0 \\ \sin\chi_a & \cos\chi_a & 0 \\ 0 & 0 & 1 \end{pmatrix} \begin{pmatrix} X_h \\ Y_h \\ Z_0 \end{pmatrix}$$

$$\begin{pmatrix} X_h \\ Y_h \\ Z_0 \end{pmatrix} = \begin{pmatrix} \cos\gamma_a & 0 & \sin\gamma_a \\ 0 & 1 & 0 \\ -\sin\gamma_a & 0 & \cos\gamma_a \end{pmatrix} \begin{pmatrix} X_a \\ Y_h \\ Z_f \end{pmatrix}$$

$$\begin{pmatrix} X_a \\ Y_h \\ Z_f \end{pmatrix} = \begin{pmatrix} 1 & 0 & 0 \\ 0 & \cos\mu_a & -\sin\mu_a \\ 0 & \sin\mu_a & \cos\mu_a \end{pmatrix} \begin{pmatrix} X_a \\ Y_a \\ Z_a \end{pmatrix}$$

$$\begin{pmatrix} X_0 \\ Y_0 \\ Z_0 \end{pmatrix} = \begin{pmatrix} \cos\chi_a * \cos\gamma_a & -\sin\chi_a & \cos\chi_a * \sin\gamma_a \\ \sin\chi_a * \cos\gamma_a & \cos\chi a & \sin\chi_a * \sin\gamma_a \\ -\sin\gamma_a & 0 & \cos\gamma_a \end{pmatrix} * \begin{pmatrix} 1 & 0 & 0 \\ 0 & \cos\mu_a & -\sin\mu_a \\ 0 & \sin\mu_a & \cos\mu_a \end{pmatrix} * \begin{pmatrix} X_a \\ Y_a \\ Z_a \end{pmatrix}$$

$$\begin{pmatrix} X_0 \\ Y_0 \\ Z_0 \end{pmatrix} = \begin{pmatrix} M_{11} & M_{12} & M_{13} \\ M_{21} & M_{22} & M_{23} \\ M_{31} & M_{32} & M_{33} \end{pmatrix} \begin{pmatrix} X_a \\ Y_a \\ Z_a \end{pmatrix}$$

$$\begin{matrix} X_0 \\ Y_0 \\ Z_0 \end{matrix} = (\text{Matrix}) \begin{pmatrix} X_a \\ Y_a \\ Z_a \end{pmatrix}$$

and:

$$\begin{pmatrix} X_a \\ Y_a \\ Z_a \end{pmatrix} = (\text{Matrix})^{-1} \begin{pmatrix} X_0 \\ Y_0 \\ Z_0 \end{pmatrix}$$

with:

$$(\text{Matrix}) = \begin{pmatrix} M_{11} & M_{12} & M_{13} \\ M_{21} & M_{22} & M_{23} \\ M_{31} & M_{32} & M_{33} \end{pmatrix}$$

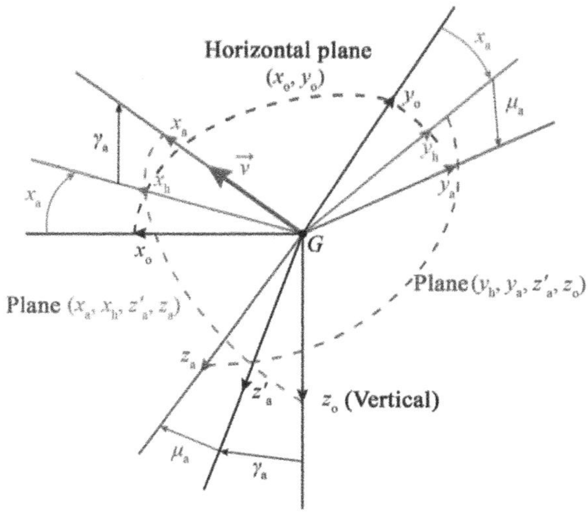

Figure 4.5. *Position of the aerodynamic trihedron with respect to the terrestrial trihedron. In blue: the terrestrial trihedron (G, x_o, y_o, z_o), Gz_o depends on the gravity; in red: the trihedron of the aerodynamic speed (G, x_a, y_a, z_a). For a color version of this figure, see www.iste.co.uk/louis/flight.zip*

The inverse matrix, which makes it possible to pass from the aircraft trihedron to the terrestrial trihedron, is symmetrical with respect to the diagonal:

$$(\text{Matrix})^{-1} = \begin{pmatrix} M_{11} & M_{21} & M_{31} \\ M_{12} & M_{22} & M_{32} \\ M_{13} & M_{23} & M_{33} \end{pmatrix}$$

where:

$$M_{11} = \cos\chi_a * \cos\gamma_a$$

$$M_{12} = -\sin\chi_a * \cos\mu_a + \cos\chi_a * \sin\gamma_a * \sin\mu_a$$

$$M_{13} = \sin\chi_a * \sin\mu_a + \cos\chi_a * \sin\gamma_a * \cos\mu_a$$

$$M_{21} = \sin\chi_a * \cos\gamma_a$$

$$M_{22} = \cos\chi_a * \cos\mu_a + \sin\chi_a * \sin\gamma_a * \sin\mu_a$$

$$M_{23} = -\cos\chi_a * \sin\mu_a + \sin\chi_a * \sin\gamma_a * \cos\mu_a$$

$$M_{31} = -\sin\gamma_a$$

$$M_{32} = \cos\gamma_a * \sin\mu_a$$

$$M_{33} = \cos\gamma_a * \cos\mu_a$$

4.5.3. *Position of the aircraft trihedron in relation to the aerodynamic speed trihedron*

The aerodynamic speed trihedron of the aircraft $(G, X_{va}, Y_{va}, Z_{va})$ is defined as follows:

– GX_{va} is driven by the aerodynamic speed;

– $GZ_{va} \perp GX_{va}$ is located in the plane of symmetry of the aircraft and oriented > 0 towards the belly of the aircraft (towards the center of the Earth);

– GY_{va} completes the direct trihedron.

Axis GZ_{va} being located by definition in the aircraft symmetry plane (G, X_e, Z_e), just two angles are enough to achieve the change of reference:

– the incidence α is the angle of the axis (GX_e) with the plane (G, X_{va}, Y_{va}) of the aerodynamic trihedron;

– the slip β is the angle of the axis (GX_{va}) with the plane of symmetry (G, X_e, Z_e).

Moreover:

– $\alpha > 0$ when (G, X_e) is above the plane;

– $\beta > 0$ if the projection of the speed vector is positive on the axis (G, Y_e).

Therefore, the coordinates (X_0, Y_0, Z_0) in the reference trihedron are written using the following transfer matrix from the coordinates (X_e, Y_e, Z_e) in the aircraft trihedron:

$$\begin{pmatrix} X_e \\ Y_e \\ Z_e \end{pmatrix} = \begin{pmatrix} \cos\alpha & 0 & -\sin\alpha \\ 0 & 1 & 0 \\ \sin\alpha & 0 & \cos\alpha \end{pmatrix} \begin{pmatrix} X_{va} \\ Y_e \\ Z_{va} \end{pmatrix}$$

$$\begin{pmatrix} X_{va} \\ Y_e \\ Z_{va} \end{pmatrix} = \begin{pmatrix} \cos\beta & -\sin\beta & 0 \\ \sin\beta & \cos\beta & 0 \\ 0 & 0 & 1 \end{pmatrix} \begin{pmatrix} X_{va} \\ Y_{va} \\ Z_{va} \end{pmatrix}$$

$$\begin{pmatrix} X_e \\ Y_e \\ Z_e \end{pmatrix} = \begin{pmatrix} \cos\alpha * \cos\beta & -\cos\alpha * \sin\beta & -\sin\alpha \\ \sin\beta & \cos\beta & 0 \\ \sin\alpha * \cos\beta & -\sin\alpha * \sin\beta & \cos\alpha \end{pmatrix} \begin{pmatrix} X_{va} \\ Y_{va} \\ Z_{va} \end{pmatrix}$$

$$\begin{pmatrix} X_e \\ Y_e \\ Z_e \end{pmatrix} = (\text{Matrix}) \begin{pmatrix} X_{va} \\ Y_{va} \\ Z_{va} \end{pmatrix}$$

and:

$$\begin{pmatrix} X_{va} \\ Y_{va} \\ Z_{va} \end{pmatrix} = (\text{Matrix})^{-1} \begin{pmatrix} X_e \\ Y_e \\ Z_e \end{pmatrix}$$

with the matrix which makes it possible to pass from the aerodynamic speed trihedron to the aircraft trihedron:

$$(\text{Matrix}) = \begin{pmatrix} \cos\alpha * \cos\beta & -\cos\alpha * \sin\beta & -\sin\alpha \\ \sin\beta & \cos\beta & 0 \\ \sin\alpha * \cos\beta & -\sin\alpha * \sin\beta & \cos\alpha \end{pmatrix}$$

The inverse matrix, which makes it possible to pass from the aircraft trihedron to the aerodynamic speed trihedron, is symmetrical with respect to the diagonal:

$$(\text{Matrix})^{-1} = \begin{pmatrix} \cos\alpha * \cos\beta & \sin\beta & \sin\alpha * \cos\beta \\ -\cos\alpha * \sin\beta & \cos\beta & -\sin\alpha * \sin\beta \\ -\sin\alpha & 0 & \cos\alpha \end{pmatrix}$$

The total angle of attack is the angle αT between axis GX_e of the aircraft and axis GX_a of the aerodynamic speed.

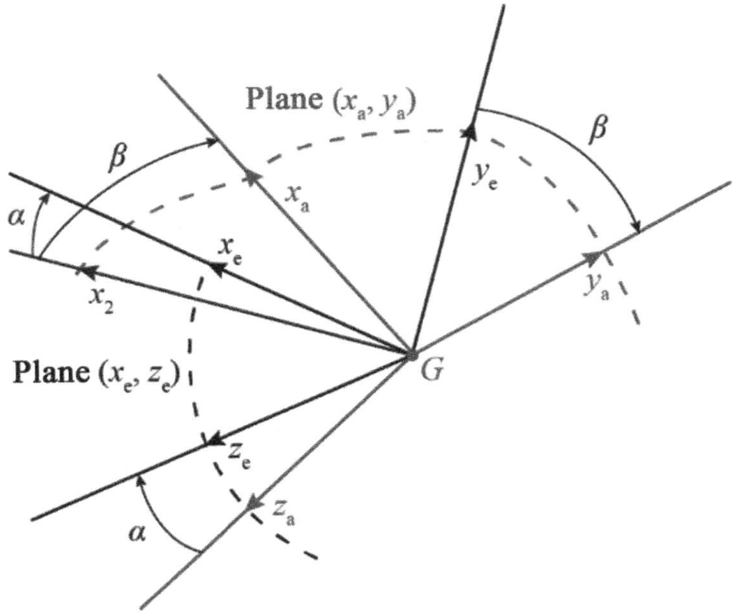

Figure 4.6. *Position of the aerodynamic trihedron with respect to the aircraft trihedron. In red: the aerodynamic speed trihedron (G, x_a, y_a, z_a); in black: the machine trihedron (G, x_e, y_e, z_e); in blue: the projection α of the axis (G, x_e) on the plane (G, x_a, y_a); in black: the projection of the axis (G, x_a) on the plane (G, x_e, z_e). Plane (G, x_a, y_a) comprises points (x_2, x_a, y_e, y_a); plane (G, x_e, z_e) comprises points (z_a, z_e, x_2, x_e). For a color version of this figure, see www.iste.co.uk/louis/flight.zip*

4.5.4. *Position of the aircraft trihedron in relation to the balance trihedron*

The position of the aircraft trihedron (G, X_e, Y_e, Z_e) with respect to the balance trihedron (G, X_b, Y_b, Z_b) is given by the equations in Figure 4.7.

Consequently, the coordinates (X_e, Y_e, Z_e) in the aircraft trihedron can be written thanks to the following transfer matrix from the coordinates (X_b, Y_b, Z_b) in the balance trihedron:

$$\begin{pmatrix} X_e \\ Y_e \\ Z_e \end{pmatrix} = \begin{pmatrix} 1 & 0 & 0 \\ 0 & \cos\phi_a & \sin\phi_a \\ 0 & -\sin\phi_a & \cos\phi_a \end{pmatrix} \begin{pmatrix} X_b \\ Y_b \\ Z_b \end{pmatrix}$$

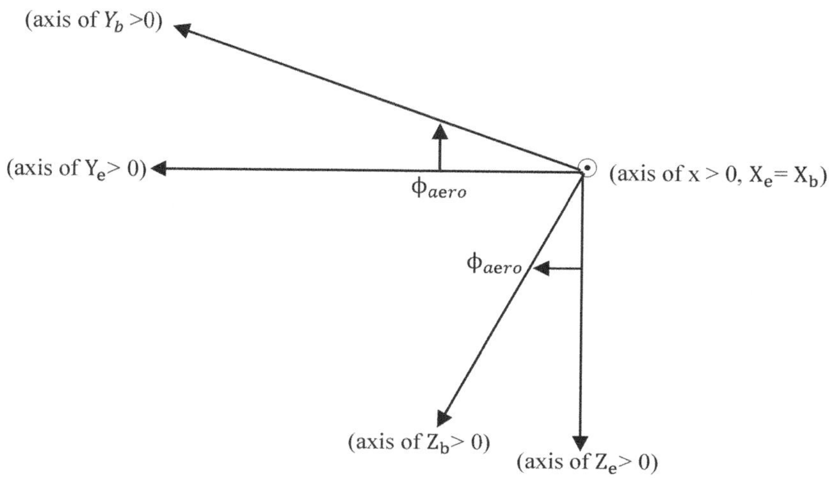

Figure 4.7. *Position of the aircraft trihedron in relation to the balance trihedron*

COMMENTS ON FIGURE 4.7.–

– $X_e = X_b$;

– $Y_e = Y_b * cos\,\phi_a - Z_b * sin\,\phi_a$;

– $Z_e = Y_b * (-\,sin\,\phi_a) + Z_b * cos\,\phi_a$.

The inverse matrix (matrix)$^{-1}$, which makes it possible to pass from the balance trihedron to the aircraft trihedron, is symmetrical with respect to the diagonal:

$$\begin{pmatrix} X_b \\ Y_b \\ Z_b \end{pmatrix} = \begin{pmatrix} 1 & 0 & 0 \\ 0 & \cos\phi_a & -\sin\phi_a \\ 0 & \sin\phi_a & \cos\phi_a \end{pmatrix} \begin{pmatrix} X_e \\ Y_e \\ Z_e \end{pmatrix}$$

The flight coefficients are given by the following tables:

$$C_{X_b} = C_{X_e}$$

$$C_{Y_b} = C_{Y_e} * \cos\phi_a - C_{Z_e} * \sin\phi_a$$

$$C_{Z_b} = C_{Y_e} * \sin\phi_a + C_{Z_e} * \cos\phi_a$$

$$C_{L_b} = C_{L_e}$$

$$C_{M_b} = C_{M_e} * \cos \phi_a - C_{N_e} * \sin \phi_a$$

$$C_{N_b} = C_{M_e} * \sin \phi_a + C_{N_e} * \cos \phi_a$$

$$C_{X_e} = C_{X_b}$$

$$C_{Y_e} = C_{Y_b} * \cos \phi_a + C_{Z_b} * \sin \phi_a$$

$$C_{Z_e} = C_{Y_b} * (-\sin \phi_a) + C_{Z_b} * \cos \phi_a$$

$$C_{L_e} = C_{L_b}$$

$$C_{M_e} = C_{M_b} * \cos \phi_a + C_{N_b} * \sin \phi_a$$

$$C_{N_e} = C_{M_b} * (-\sin \phi_a) + C_{N_b} * \cos \phi_a$$

The aerodynamic coefficients obtained in the wind tunnel are the coefficients generally defined in [0°, 45°].

4.5.5. Position of the terrestrial trihedron in relation to the local geographic trihedron

The meanings of the parameters used are as follows:

– Ω: the speed of rotation of the Earth with respect to an absolute reference;

– λ_M: the difference between the latitude of the aircraft and the latitude of the origin of the test track;

– λ_{ct}: the latitude of the origin of the test track;

– G_M: the difference between the longitude of the aircraft and the longitude of the origin of the test track.

The position of the aircraft in the land reference is given by the following transfer matrix using two successive rotations. The axis (GZ_{lgt}) being located by definition in the aircraft plane of symmetry (G, X_e, Z_e), just two angles are enough to achieve the change of reference:

– the latitude $(\lambda_M + \lambda_{ct})$ is the angle of the axis (GX_e) reduced to (GX_{lgt}) with the plane (G, X_{ter}, Y_{ter}) of the aerodynamic trihedron;

– the longitude (G_M) is the angle of the axis (GX_e) reduced to (GX_{lgt}) with the plane of symmetry (G, X_{ter}, Y_{ter}).

When (GX_{lgt}) is above the plane, $(\lambda_M + \lambda_{ct}) > 0$. If the projection of the speed vector is positive on the axis (GY_{lgt}), $(G_M) > 0$.

Therefore, the coordinates $(X_{ter}, Y_{ter}, Z_{ter})$ d in the terrestrial trihedron are written using the following transfer matrix from the coordinates $(X_{lgt}, Y_{lgt}, Z_{lgt})$ in the local geographic trihedron:

$$\begin{pmatrix} X_{ter} \\ Y_{ter} \\ Z_{ter} \end{pmatrix} = \begin{pmatrix} \cos(\lambda_M + \lambda_{ct}) & 0 & -\sin(\lambda_M + \lambda_{ct}) \\ 0 & 1 & 0 \\ \sin(\lambda_M + \lambda_{ct}) & 0 & \cos(\lambda_M + \lambda_{ct}) \end{pmatrix} \begin{pmatrix} X_{lgt} \\ Y_{ter} \\ Z_{lgt} \end{pmatrix}$$

$$\begin{pmatrix} X_{lgt} \\ Y_{ter} \\ Z_{lgt} \end{pmatrix} = \begin{pmatrix} \cos G_M & -\sin G_M & 0 \\ \sin G_M & \cos G_M & 0 \\ 0 & 0 & 1 \end{pmatrix} \begin{pmatrix} X_{lgt} \\ Y_{lgt} \\ Z_{lgt} \end{pmatrix}$$

$$\begin{pmatrix} X_{ter} \\ Y_{ter} \\ Z_{ter} \end{pmatrix} =$$

$$\begin{pmatrix} \cos(\lambda_M + \lambda_{ct}) * \cos G_M & -\cos(\lambda_M + \lambda_{ct}) * \sin G_M & -\sin(\lambda_M + \lambda_{ct}) \\ \sin G_M & \cos G_M & 0 \\ \sin(\lambda_M + \lambda_{ct}) * \cos G_M & -\sin(\lambda_M + \lambda_{ct}) * \sin G_M & \cos(\lambda_M + \lambda_{ct}) \end{pmatrix} \begin{pmatrix} X_{lgt} \\ Y_{lgt} \\ Z_{lgt} \end{pmatrix}$$

The matrix that makes it possible to pass from the terrestrial trihedron to the local geographical trihedron is therefore:

$$\begin{pmatrix} \cos(\lambda_M + \lambda_{ct}) * \cos G_M & -\cos(\lambda_M + \lambda_{ct}) * \sin G_M & -\sin(\lambda_M + \lambda_{ct}) \\ \sin G_M & \cos G_M & 0 \\ \sin(\lambda_M + \lambda_{ct}) * \cos G_M & -\sin(\lambda_M + \lambda_{ct}) * \sin G_M & \cos(\lambda_M + \lambda_{ct}) \end{pmatrix}$$

The inverse matrix that makes it possible to pass from the local geographical trihedron to the terrestrial trihedron is symmetrical relative to the diagonal:

$$\begin{pmatrix} \cos(\lambda_M + \lambda_{ct}) * \cos G_M & \sin G_M & \sin(\lambda_M + \lambda_{ct}) * \cos G_M \\ -\cos(\lambda_M + \lambda_{ct}) * \sin G_M & \cos G_M & -\sin(\lambda_M + \lambda_{ct}) * \sin G_M \\ -\sin(\lambda_M + \lambda_{ct}) & 0 & \cos(\lambda_M + \lambda_{ct}) \end{pmatrix}$$

5

Movement of a Solid

5.1. Rotation of a solid around a fixed point

Consider a solid (S) mobile around a fixed point O: a point of the solid permanently corresponds with the point O.

Let (O, I_1, J_1, K_1) be a frame of reference (R_1) rigidly linked to (S) and (O, I, J, K) a frame of reference (R) with respect to which we propose to study the motion of (S).

We can transform (R) into (R1) with three successive rotations:

– (R) $(O, I, J, K) \rightarrow (R')$ (O, U, V, K) by rotation Ψ around $(\overrightarrow{OK})$;

– (R') $(O, U, V, K) \rightarrow (R'')$ (O, U, V', K_1) by rotation θ around $(\overrightarrow{OU})$;

– (R'') $(O, U, V', K_1) \rightarrow (R_1)$ (O, I_1, J_1, K_1) by rotation φ around (O, K1).

The speed vector of a point of (S) is zero with respect to (R1), and with respect to (R"), it is:

$$\vec{V}_{r''}(M) = \left(\frac{\partial \phi}{\partial t}\right) * \vec{K}_1 \wedge \overrightarrow{OM}$$

In R', we must add the drive speed of a point of (R") with respect to (R'):

$$\vec{V}_{r''}(M) = \left(\frac{\partial \phi}{\partial t}\right) * \vec{K}_1 \wedge \overrightarrow{OM} + \left(\frac{\partial \theta}{\partial t}\right) * \vec{U} \wedge \overrightarrow{OM}$$

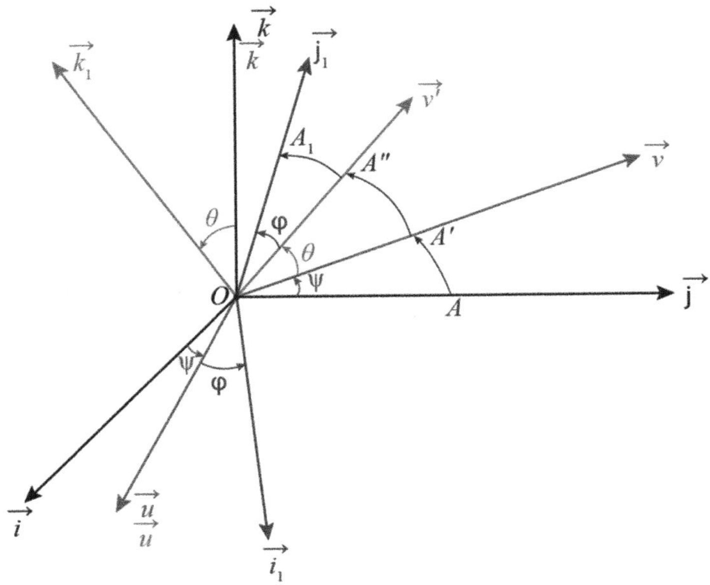

Figure 5.1. *Rotation of a solid around a fixed point. For a color version of this figure, see www.iste.co.uk/louis/flight.zip*

Likewise, we obtain the velocity vector of M with respect to (R):

$$\vec{V}_{r''}(M) = \left(\frac{\partial \phi}{\partial t}\right) * \vec{K}_1 \wedge \overrightarrow{OM} + \left(\frac{\partial \theta}{\partial t}\right) * \vec{U} \wedge \overrightarrow{OM} + \left(\frac{\partial \Psi}{\partial t}\right) * \vec{K} \wedge \overrightarrow{OM}$$

with:

$$\vec{V}_r(M) = \vec{\Omega} \wedge \overrightarrow{OM}$$

by letting:

$$\vec{\Omega} = \left(\frac{\partial \phi}{\partial t}\right) * \vec{K}_1 + \left(\frac{\partial \theta}{\partial t}\right) * \vec{U} + \left(\frac{\partial \Psi}{\partial t}\right) * \vec{K}$$

where $\vec{\Omega}$ is the speed vector of rotation of the solid (S) in its motion with respect to (R).

Ψ, θ and φ are the Euler angles, with:

– Ψ: the precession angle;

– Θ: the nutation angle;

– Φ: proper angle of rotation.

The rotation of angle φ around K1 translates analytically as:

$$\begin{pmatrix} U \\ V' \\ K_1 \end{pmatrix} = \begin{pmatrix} \cos\phi & -\sin\phi & 0 \\ \sin\phi & \cos\phi & 0 \\ 0 & 0 & 1 \end{pmatrix} \begin{pmatrix} I_1 \\ J_1 \\ K_1 \end{pmatrix}$$

The rotation of angle θ around U translates analytically into:

$$\begin{pmatrix} U \\ V \\ K \end{pmatrix} = \begin{pmatrix} 1 & 0 & 0 \\ 0 & \cos\theta & -\sin\theta \\ 0 & \sin\theta & \cos\theta \end{pmatrix} \begin{pmatrix} U \\ V' \\ K_1 \end{pmatrix}$$

From this we obtain:

$$\vec{U} = \cos\phi * \vec{I}_1 - \sin\phi * \vec{J}_1$$

$$\vec{K} = \sin\theta * \vec{V}' + \cos\theta * \vec{K}_1$$
$$= \sin\theta * (\sin\phi * \vec{I}_1 + \cos\phi * \vec{J}_1) + \cos\theta * \vec{K}_1$$

$$\vec{K} = \sin\theta * \sin\phi * \vec{I}_1 + \sin\theta * \cos\phi * \vec{J}_1 + \cos\theta * \vec{K}_1$$

with:

$$\vec{\Omega} = \left(\frac{\partial\Phi}{\partial t}\right) * \vec{K}_1 + \left(\frac{\partial\theta}{\partial t}\right) * \vec{U} + \left(\frac{\partial\Psi}{\partial t}\right) * \vec{K}$$

$$\vec{\Omega} = \left(\frac{\partial\phi}{\partial t}\right) * \vec{K}_1 + \left(\frac{\partial\theta}{\partial t}\right) * (\cos\phi * \vec{I}_1 - \sin\phi * \vec{J}_1) + \left(\frac{\partial\Psi}{\partial t}\right)$$
$$* (\sin\theta * \sin\phi * \vec{I}_1 + \sin\theta * \cos\phi * \vec{J}_1 + \cos\theta * \vec{K}_1)$$

The components p_1, q_1 and r_1 of $\vec{\Omega}$ on $\vec{I}_1, \vec{J}_1, \vec{K}_1$ are therefore:

$$p_1 = \left(\frac{\partial\theta}{\partial t}\right) * \cos\phi + \left(\frac{\partial\Psi}{\partial t}\right) * \sin\theta * \sin\phi \text{ on axis } \vec{I}_1$$

$$q_1 = -\left(\frac{\partial\theta}{\partial t}\right) * \sin\phi + \left(\frac{\partial\Psi}{\partial t}\right) * \sin\theta * \cos\phi \text{ on axis } \vec{J}_1$$

$$r_1 = \left(\frac{\partial\phi}{\partial t}\right) + \left(\frac{\partial\Psi}{\partial t}\right) * \cos\theta * \vec{K}_1 \text{on axis } \vec{K}_1$$

The rotation of angle Ψ around K translates analytically in the base (O, I, J, K) into:

$$\begin{pmatrix} U \\ V \\ K \end{pmatrix} = \begin{pmatrix} \cos\Psi & \sin\Psi & 0 \\ -\sin\Psi & \cos\Psi & 0 \\ 0 & 0 & 1 \end{pmatrix} \begin{pmatrix} I \\ J \\ K \end{pmatrix}$$

The rotation of angle θ around U translates analytically into:

$$\begin{pmatrix} U \\ V' \\ K_1 \end{pmatrix} = \begin{pmatrix} 1 & 0 & 0 \\ 0 & \cos\theta & \sin\theta \\ 0 & -\sin\theta & \cos\theta \end{pmatrix} \begin{pmatrix} U \\ V \\ K \end{pmatrix}$$

from which we obtain:

$$\vec{U} = \cos\Psi * \vec{I} + \sin\Psi * \vec{J}$$

$$\vec{V} = -\sin\Psi * \vec{I} + \cos\Psi * \vec{J}$$

$$\vec{K}_1 = -\sin\theta * \vec{V} + \cos\theta * \vec{K}$$

$$\vec{K}_1 = -\sin\theta * (-\sin\Psi * \vec{I} + \cos\Psi * \vec{J}) + \cos\theta * \vec{K}$$

with:

$$\vec{\Omega} = \left(\frac{\partial\phi}{\partial t}\right) * \vec{K}_1 + \left(\frac{\partial\theta}{\partial t}\right) * \vec{U} + \left(\frac{\partial\Psi}{\partial t}\right) * \vec{K}$$

and:

$$\vec{\Omega} = \left(\frac{\partial\phi}{\partial t}\right) * \left(-\sin\theta * (-\sin\Psi * \vec{I} + \cos\Psi * \vec{J}) + \cos\theta * \vec{K}\right) + \left(\frac{\partial\theta}{\partial t}\right) * \left(\cos\Psi * \vec{I} + \sin\Psi * \vec{J}\right) + \left(\frac{\partial\Psi}{\partial t}\right) * \vec{K}$$

The components p, q and r of $\vec{\Omega}$ on $\vec{I}, \vec{J}, \vec{K}$ are therefore:

$$p = \left(\frac{\partial \theta}{\partial t}\right) * \cos \Psi + \left(\frac{\partial \Phi}{\partial t}\right) * \sin \theta * \text{ on axis } \vec{I}$$

$$q = \left(\frac{\partial \theta}{\partial t}\right) * \sin \Psi - \left(\frac{\partial \Phi}{\partial t}\right) * \sin \theta * \cos \Psi \text{ on axis } \vec{J}$$

$$r = \left(\frac{\partial \Phi}{\partial t}\right) * \cos \theta + \left(\frac{\partial \Psi}{\partial t}\right) \text{ on axis } \vec{K}$$

5.2. Kinematics of a solid

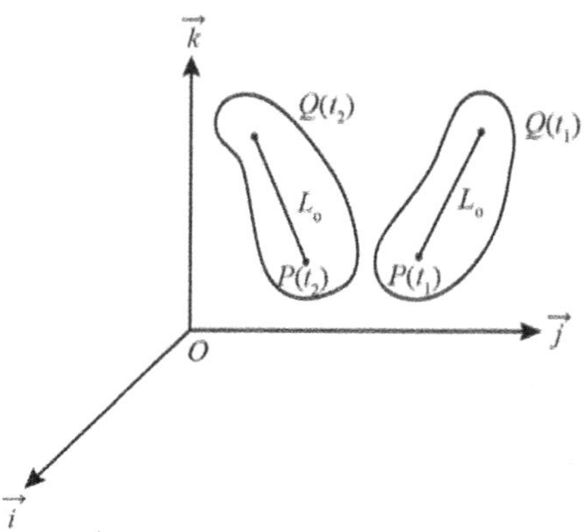

Figure 5.2. *Kinematics of a solid*

If P and Q are the positions of two particles of a solid at time t, the modulus of PQ is independent of time:

$$|PQ|^2 = L_0^2$$

Let P_1, Q_1, P_2 and Q_2 be the positions of four particles at time t with:

$$L_1 = P_1Q_1 \text{ and } L_2 = P_2Q_2$$

The scalar product $L_1 . L_2$ is independent of time.

The solid velocity field is given by derivation of the scalar product $(L_1 . L_2)$:

$$L_1 * \left(\frac{\partial L_2}{\partial t}\right) + \left(\frac{\partial L_1}{\partial t}\right) * L_2 = 0 \Rightarrow L_1 * \left(\frac{\partial L_2}{\partial t}\right) = -\left(\frac{\partial L_1}{\partial t}\right) * L_2$$

The derivation with respect to time thus appears as a linear antisymmetric map, which can be described as:

$$\left(\frac{\partial L}{\partial t}\right) = \vec{\Omega} \wedge \vec{L}$$

$\vec{\Omega}$ is the vector of the antisymmetric map and in this case is called the rotation vector.

From:

$$L = PQ = OQ - OP$$

we obtain:

$$\overrightarrow{V(Q)} - \overrightarrow{V(P)} = \left(\frac{\partial \vec{L}}{\partial t}\right) + \left(\frac{\partial \overrightarrow{OQ}}{\partial t}\right) + \left(\frac{\partial \overrightarrow{OP}}{\partial t}\right)$$

and therefore:

$$\overrightarrow{V(Q)} = \overrightarrow{V(P)} + \vec{\Omega} \wedge \vec{L} = \overrightarrow{V(P)} + \vec{\Omega} \wedge \overrightarrow{PQ}$$

Acceleration field of a solid

By deriving with respect to time the relation, which gives $\overrightarrow{V(Q)}$, we obtain:

$$\left(\frac{\partial \overrightarrow{V(Q)}}{\partial t}\right) = \left(\frac{\partial \overrightarrow{V(P)}}{\partial t}\right) + \left(\frac{\partial \vec{\Omega}}{\partial t}\right) \wedge \overrightarrow{PQ} + \vec{\Omega} \wedge \left(\frac{\partial \overrightarrow{PQ}}{\partial t}\right)$$

or:

$$\overrightarrow{\gamma(Q)} = \overrightarrow{\gamma(P)} + \left(\frac{\partial\vec{\Omega}}{\partial t}\right) \wedge \overrightarrow{PQ} + \vec{\Omega} \wedge \left(\frac{\partial\overrightarrow{PQ}}{\partial t}\right)$$

$$\overrightarrow{\gamma(Q)} = \overrightarrow{\gamma(P)} + \left(\frac{\partial\vec{\Omega}}{\partial t}\right) \wedge \overrightarrow{PQ} + \vec{\Omega} \wedge \left(\frac{\partial\overrightarrow{OQ}}{\partial t}\right) - \vec{\Omega} \wedge \left(\frac{\partial\overrightarrow{OP}}{\partial t}\right)$$

or:

$$\overrightarrow{\gamma(Q)} = \overrightarrow{\gamma(P)} + \left(\frac{\partial\vec{\Omega}}{\partial t}\right) \wedge \overrightarrow{PQ} + \vec{\Omega} \wedge (\vec{\Omega} \wedge \overrightarrow{PQ})$$

Starting from the following mathematical identity:

$$\vec{U} \wedge (\vec{V} \wedge \vec{W}) = (\vec{U} * \vec{W}) * \vec{V} - (\vec{U} * \vec{V}) * \vec{W}$$

$\vec{U} \wedge \vec{V}$ is written:

$$\begin{pmatrix} U1 \\ U2 \\ U3 \end{pmatrix} \wedge \begin{pmatrix} V1 \\ V2 \\ V3 \end{pmatrix} = \begin{pmatrix} U2 * V3 - V2 * U3 \\ U3 * V1 - V3 * U1 \\ U1 * V2 - V1 * U2 \end{pmatrix}$$

and $\vec{U} \wedge (\vec{V} \wedge \vec{W})$ is written:

$$\begin{pmatrix} U1 \\ U2 \\ U3 \end{pmatrix} \wedge \begin{pmatrix} V2 * U3 - V3 * U2 \\ V3 * W1 - V1 * W3 \\ V1 * W2 - V2 * W1 \end{pmatrix}$$
$$= \begin{pmatrix} (U2W2 + U3W3)\,V1 - (U2V2 + U3V3)\,W1 \\ (U3W3 + U1W1)\,V2 - (U3V3 + U1V1)\,W2 \\ (U1W1 + U2W2)\,V3 - (U1V1 + U2V2)\,W3 \end{pmatrix}$$

Now:

$$\overrightarrow{V(Q)} - \overrightarrow{V(P)} = \vec{\Omega} \wedge \overrightarrow{PQ}$$

$$\vec{\Omega} \wedge (\vec{\Omega} \wedge \overrightarrow{PQ}) = (\vec{\Omega} * \overrightarrow{PQ}) * \vec{\Omega} - \vec{\Omega} * \vec{\Omega} * \overrightarrow{PQ}$$

giving the relation:

$$\overrightarrow{\gamma(Q)} = \overrightarrow{\gamma(P)} + \left(\frac{\partial\vec{\Omega}}{\partial t}\right) \wedge \overrightarrow{PQ} + \vec{\Omega} \wedge (\vec{\Omega} \wedge \overrightarrow{PQ})$$

or, by development:

$$\overrightarrow{\gamma(Q)} = \overrightarrow{\gamma(P)} + \left(\frac{\partial\vec{\Omega}}{\partial t}\right) \wedge \overrightarrow{PQ} + ((\vec{\Omega} * \overrightarrow{PQ}) * \vec{\Omega} - \vec{\Omega} * \vec{\Omega} * \overrightarrow{PQ})$$

5.3. Reference framework

5.3.1. *Absolute reference*

Let there be a material point M observed with respect to a reference point (R_A) of center O,$(O, \vec{\imath}, \vec{\jmath}, \vec{k})$:

$$\overrightarrow{OM} = x * \vec{\imath} + y * \vec{\jmath} + \vec{k} \quad [5.1]$$

We can say that this first frame of reference (R_A) makes it possible to observe the absolute movement of the mobile point M. The only privilege of this reference point is to have been chosen as a reference and, *a priori*, it does not present any particular property; we call it the absolute reference. Its vectors $\vec{\imath}, \vec{\jmath}$ and $\vec{k}$ will be considered constant.

5.3.2. *Relative reference*

Let there be a second reference (R_R) with a P center $(P, \vec{I}, \vec{J}, \vec{K})$ called the relative reference, which is in unspecified movement compared to the absolute reference.

The point M will be identified in this relative reference by:

$$\overrightarrow{PM} = X * \vec{I} + Y * \vec{J} + Z * \vec{K} \quad [5.2]$$

Its vectors $\vec{I}, \vec{J}, \vec{K}$ are not fixed in the absolute coordinate system (R_A).

The movement of M observed in this second reference point (R_R) is said to be relative movement.

At a given instant t, this movement of (R_R) with respect to (R_A) is characterized by two quantities:

– the position of point P, origin of the coordinate system (P, X, Y, Z):

$$\overrightarrow{OP} = xp * \vec{\imath} + yp * \vec{\jmath} + zp * \vec{k} \quad [5.3]$$

– a rotation of (R_R) that we will define by a rotation vector $\Omega_{R/A}$ (note that Ω is rarely constant, especially in modulus or direction, even for the Earth).

5.3.3. *Drive movement*

A fixed point in the relative reference (X, Y and Z constants) is a mobile point in the absolute reference. The movement of such a point in the absolute frame of reference is said to be a drive movement.

5.3.4. *Goal*

We will try to express the position, the speed and the acceleration in the reference (R_A) $(O, \vec{\imath}, \vec{\jmath}, \vec{k})$ according to:

– their counterparts in the relative frame of reference (R_R);

– the characteristics of the movement of the frame (R_R) with respect to the absolute reference ($R_{A''}$).

5.3.5. *Change of reference framework*

In Galilean kinematics, or classical kinematics, the observers linked to the two reference frames perceive an event as taking place at the same point of reference ξ (ξ affine space).

The association of the space reference $(O, \vec{I}, \vec{J}, \vec{K})$ and the time reference constitutes a reference (R). The association of the space reference (O_1, I_1, J_1, K_1) and the time reference constitutes a reference (R_1) (Bruhat 1967; Annequin and Boutigny 1974).

5.3.5.1. *The movement of (R1) with respect to (R)*

The reference $(O, \vec{I}, \vec{J}, \vec{K})$ is undeformable over time and as such can be considered as a solid.

Each of the unit vectors maintains a constant modulus. We can apply to each of them the relations of the kinematics of a solid.

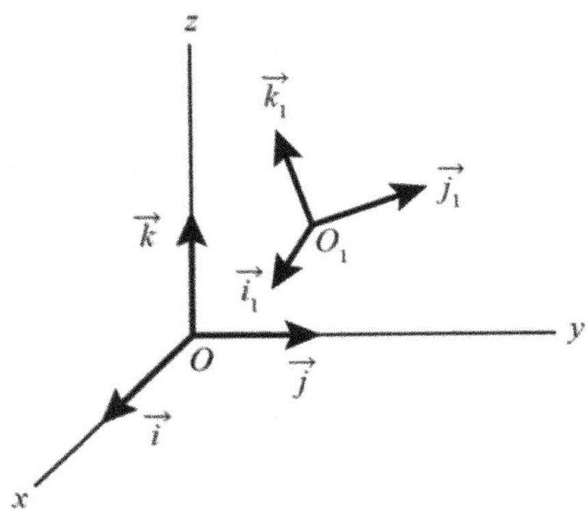

Figure 5.3. *Change of reference framework*

COMMENTS on FIGURE 5.3.–

$$\left(\frac{\partial \vec{I1}}{\partial t}\right) = \vec{\Omega}_e \wedge \vec{I}_1$$

$$\left(\frac{\partial \vec{J1}}{\partial t}\right) = \vec{\Omega}_e \wedge \vec{J}_1$$

$$\left(\frac{\partial \vec{K1}}{\partial t}\right) = \vec{\Omega}_e \wedge K_1$$

Ω_e is the rotation vector of (R_1) with respect to the reference (R).

5.3.5.2. *Derivation composed of a vector*

Let there be a vector K whose components in the reference (R) are α, β and γ on $(\vec{I}, \vec{J}, \vec{K})$ and α_1, β_1 and γ_1 in the reference (R_1).

$$\vec{K} = \alpha * \vec{I} + \beta * \vec{J} + \gamma * \vec{K}$$

$$\vec{K} = \alpha_1 * \vec{I}_1 + \beta_1 * \vec{J}_1 + \gamma_1 * \vec{K}_1$$

The vector derived from K with respect to time in the reference (R) is given by:

$$\left(\frac{\partial \vec{K}}{\partial t}\right)_{(R)} = \left(\frac{\partial \alpha}{\partial t}\right) * \vec{I} + \left(\frac{\partial \beta}{\partial t}\right)\vec{J} + \left(\frac{\partial \gamma}{\partial t}\right)\vec{K}$$

The vector derived from K with respect to time in the reference (R_1) is given by:

$$\left(\frac{\partial \vec{K}}{\partial t}\right)_{(R_1)} = \left(\frac{\partial \alpha_1}{\partial t}\right) * \vec{I}_1 + \left(\frac{\partial \beta_1}{\partial t}\right) * \vec{J}_1 + \left(\frac{\partial \gamma_1}{\partial t}\right)\vec{K}_1$$

We can get the vector derived from K in (R) from:

$$\vec{K} = \alpha_1 * \vec{I}_1 + \beta_1 * \vec{J}_1 + \gamma_1 * \vec{K}_1$$

where we consider $(\vec{I}_1, \vec{J}_1, \vec{K}_1)$ as functions of time.

$$\left(\frac{\partial \vec{K}}{\partial t}\right)_{(R)} = \left(\frac{\partial \alpha_1}{\partial t}\right) * \vec{I}_1 + \left(\frac{\partial \vec{I}_1}{\partial t}\right) * \alpha_1 + \left(\frac{\partial \beta_1}{\partial t}\right) * \vec{J}_1 + \left(\frac{\partial \vec{J}_1}{\partial t}\right) * \beta_1 + \left(\frac{\partial \gamma_1}{\partial t}\right) * \vec{K}_1 + \left(\frac{\partial \vec{K}_1}{\partial t}\right) * \gamma_1$$

$$\left(\frac{\partial \vec{K}}{\partial t}\right)_{(R)} = \left(\frac{\partial \vec{K}}{\partial t}\right)_{(R_1)} + \left(\frac{\partial \vec{I}_1}{\partial t}\right) * \alpha_1 + \left(\frac{\partial \vec{J}_1}{\partial t}\right) * \beta_1 + \left(\frac{\partial \vec{K}_1}{\partial t}\right) * \gamma_1$$

$$\left(\frac{\partial \vec{K}}{\partial t}\right)_{(R)} = \left(\frac{\partial \vec{K}}{\partial t}\right)_{(R_1)} + \vec{\Omega}_e \wedge \vec{K}$$

5.3.5.3. *Velocity vector of a point with respect to (R) and to (R_1)*

Let M be a point with coordinates (x, y, z) in (R) and in (R_1).

$$\overrightarrow{OM} = x * \vec{I} + y * \vec{J} + z * \vec{K}$$

$$\overrightarrow{O_1M} = x_1 * \vec{I}_1 + y_1 * \vec{J}_1 + z_1 * \vec{K}_1$$

The vectors derived from M in the frame (R) and in the reference (R_1) are:

$$\vec{V}_R(M) = \left(\frac{\partial \overrightarrow{OM}}{\partial t}\right)_{(R)} = \left(\frac{\partial x}{\partial t}\right) * \vec{I} + \left(\frac{\partial y}{\partial t}\right) * \vec{J} + \left(\frac{\partial z}{\partial t}\right) * \vec{K}$$

The vector derived from K in the reference (R_1) is:

$$\vec{V}_{R1}(M) = \left(\frac{\partial \overrightarrow{O1M}}{\partial t}\right)_{(R1)} = \left(\frac{\partial x1}{\partial t}\right) * \vec{I}_1 + \left(\frac{\partial y1}{\partial t}\right) * \vec{J}_1 + \left(\frac{\partial z1}{\partial t}\right) * \vec{K}_1$$

From $\overrightarrow{OM} = \overrightarrow{OO1} + \overrightarrow{O1M}$, we obtain:

$$\vec{V}_R(M) = \left(\frac{\partial \overrightarrow{OO1}}{\partial t}\right)_{(R)} + \left(\frac{\partial \overrightarrow{O1M}}{\partial t}\right)_{(R1)}$$

By applying the above formula:

$$\left(\frac{\partial \vec{K}}{\partial t}\right)_{(R)} = \left(\frac{\partial \vec{K}}{\partial t}\right)_{(R1)} + \overrightarrow{\Omega e} \wedge \vec{K}$$

we obtain:

$$\vec{V}_R(M) = \vec{V}_R(O_1) + \vec{\Omega}_e \wedge \overrightarrow{O_1M} + \vec{V}_{R1}(O_1)$$

The velocity vector of the point M considered to be linked to (R_1) is by definition the drive speed vector:

$$\vec{V}_e(M) = \vec{V}_R(O_1) + \vec{\Omega}_e \wedge \overrightarrow{O_1M}$$

$$\Rightarrow \vec{V}_R(M) = \vec{V}_e(M) + \vec{V}_{R1}(O_1)$$

5.3.5.4. *Acceleration vector of a point with respect to (R) and to (R_1)*

Let M be a coordinate point (x, y, z) in (R) and (x_1, y_1, z_1) in (R_1). The acceleration vectors of M with respect to (R) and to (R_1) are given by the following equations:

$$\vec{\gamma}_R(M) = \left(\frac{\partial \vec{V}_R(M)}{\partial t}\right)_{(R)} = \frac{\partial^2 x}{\partial t^2} * \vec{I} + \frac{\partial^2 y}{\partial t^2} * \vec{J} + \frac{\partial^2 z}{\partial t^2} * \vec{K}$$

The vector derived from $\vec{K}$ in the reference (R_1) is:

$$\vec{\gamma}_{R1}(M) = \left(\frac{\partial \vec{V}_R(M)}{\partial t}\right)_{(R1)} = \frac{\partial^2 x_1}{\partial t^2} * \vec{I}_1 + \frac{\partial^2 y_1}{\partial t^2} * \vec{J}_1 + \frac{\partial^2 z_1}{\partial t^2} * \vec{K}_1$$

By deriving:

$$\vec{V}_R(M) = \vec{V}_R(O_1) + \vec{\Omega}_e \wedge \overrightarrow{O_1M} + \vec{V}_{R1}(O_1)$$

we obtain:

$$\vec{\gamma}_R(M) = \left(\frac{\partial \vec{V}_R(M)}{\partial t}\right)_{(R)}$$

$$\vec{\gamma}_R(M) = \vec{\gamma}_R(O_1) + \left(\frac{\partial \vec{\Omega}_e}{\partial t}\right) \wedge \overrightarrow{O_1M} + \vec{\Omega}_e \wedge \left(\frac{\partial \overrightarrow{O_1M}}{\partial t}\right)_{(R)} + \vec{\Omega} \wedge \left(\frac{\partial \vec{V}_{R1}(M)}{\partial t}\right)(R)$$

$$\vec{\gamma}_R(M) = \vec{\gamma}_R(O_1) + \left(\frac{\partial \vec{\Omega}_e}{\partial t}\right) \wedge \overrightarrow{O_1M} + \left(\frac{\partial \vec{V}_{R1}(M)}{\partial t}\right)_{(R1)} + \vec{\Omega}_e \wedge \vec{V}_{R1}(M) + \vec{\Omega}_e \wedge \left(\frac{\partial \overrightarrow{O_1M}}{\partial t}\right)_{(R1)} + \vec{\Omega}_e \wedge \overrightarrow{O_1M}$$

because:

$$\left(\frac{\partial \overrightarrow{O_1M}}{\partial t}\right)_{(R)} = \left(\frac{\partial \overrightarrow{O_1M}}{\partial t}\right)_{(R1)} + \vec{\Omega}_e \wedge \overrightarrow{O_1M} = \vec{V}_{R1}(M) + \vec{\Omega}_e \wedge \overrightarrow{O_1M}$$

and:

$$\left(\frac{\partial \vec{V}_{R1}(M)}{\partial t}\right)_{(R)} = \left(\frac{\partial \vec{V}_{R1}(M))}{\partial t}\right)_{(R1)} + \vec{\Omega}_e \wedge \vec{V}_{R1}(M)$$

$\rightarrow$

$$\vec{\gamma}_R(M) = \vec{\gamma}_R(O_1) + \left(\frac{\partial \vec{\Omega}_e}{\partial t}\right) \wedge \overrightarrow{O_1M} + \vec{\gamma}_{R1}(M) + \vec{\Omega}_e \wedge \vec{V}_{R1}(M) + \vec{\Omega}_e \wedge \vec{V}_{R1}(M) + \vec{\Omega}_e \wedge \overrightarrow{O_1M}$$

$$\vec{\gamma}_R(M) = \vec{\gamma}_R(O_1) + \left(\frac{\partial \vec{\Omega}_e}{\partial t}\right) \wedge \overrightarrow{O_1M} + \vec{\gamma}_{R1}(M) + \vec{\Omega}_e \wedge \vec{V}_{R1}(M) + \vec{\Omega}_e \wedge \vec{V}_{R1}(M) + \vec{\Omega}_e \wedge (\vec{\Omega}_e \wedge \overrightarrow{O_1M})$$

$$\vec{\gamma}_R(M) = \vec{\gamma}_R(O_1) + \left(\frac{\partial \vec{\Omega}_e}{\partial t}\right) \wedge \overrightarrow{O_1M} + \vec{\Omega}_e \wedge (\vec{\Omega}_e \wedge \overrightarrow{O_1M}) + 2 * \vec{\Omega}_e \wedge \vec{V}_{R1}(M) \ + \vec{\gamma}_{R1}(M)$$

The three following terms constitute the acceleration vector in the reference (R) of the point M considered as fixed in the reference (R_1).

It is by definition the acceleration vector of M, denoted as $\vec{\gamma}_e(M)$:

$$\vec{\gamma}_e(M) = \vec{\gamma}_R(O_1) + \left(\frac{\partial \vec{\Omega}_e}{\partial t}\right) \wedge \overrightarrow{O_1M} + \vec{\Omega}_e \wedge (\vec{\Omega}_e \wedge \overrightarrow{O_1M})$$

The complementary term $2 * \vec{\Omega}_e \wedge \vec{V}_{R1}(M)$ is the complementary acceleration vector $\overrightarrow{\gamma c(M)}$ or the Coriolis force:

$$\vec{\gamma}_c(M) = 2 * \vec{\Omega}_e \wedge \vec{V}_{R1}(M)$$

Finally, the term $\vec{\gamma}_{R1}(M)$ constitutes the acceleration vector in the reference (R) of the reference (R_1).

The acceleration vector of a point with respect to (R) and to (R_1) can be written as follows:

$$\vec{\gamma}_R(M) = \vec{\gamma}_e(M) + \vec{\gamma}_c(M) + \vec{\gamma}_{R1}(M)$$

5.3.6. *Change of reference: conclusion and summary*

Some definitions:

– V_r is the relative speed, i.e. the speed of the point M in the relative reference frame (R_R);

– V_e is the speed of the coincidence point P in the absolute reference (R_A);

– Γ_a is the absolute acceleration, i.e. the acceleration of point M in the absolute reference (R_A);

– Γ_r is the relative acceleration, i.e. the acceleration of the point M in the relative reference (R_R);

– Γ_e is the acceleration of the coincidence point P in the absolute reference (R_A).

DEFINITION OF A COINCIDENCE POINT.– The relative reference point which, at date t, coincides with the moving point M (e.g. a bicycle, insect on a clock).

Knowledge of reference changes is not strictly necessary. It is obviously perfectly possible to calculate acceleration directly without using the relations established above. However, by decomposing the problem correctly, these relations often allow the necessary vector quantities to be expressed more quickly.

Remember that the demonstrations were carried out in Cartesian coordinates for reasons of convenience. It is obvious that, in the following summary, the Cartesian components of the vectors $\overrightarrow{OP}$, $\overrightarrow{PM}$, $\overrightarrow{V_p}$, $\overrightarrow{V_r}$, $\overrightarrow{\Gamma_p}$ and $\overrightarrow{\Gamma_r}$ can be replaced by their counterparts in cylindrical or spherical coordinates.

Position

$$\overrightarrow{OM} = \overrightarrow{OP} + \overrightarrow{PM}$$

$$\overrightarrow{OP} = x_p * \vec{i} + y_p * \vec{j} + z_p * \vec{k}$$

$$\overrightarrow{PM} = X * \vec{I} + Y * \vec{J} + Z * \vec{K}$$

Velocity

$$\overrightarrow{V_a} = \overrightarrow{V_p} + \vec{\Omega} \wedge \overrightarrow{PM} + \overrightarrow{V_r}$$

where:

$$\overrightarrow{V_p} = \left(\frac{\partial x_p}{\partial t}\right) * \vec{i} + \left(\frac{\partial y_p}{\partial t}\right) * \vec{j} + \left(\frac{\partial z_p}{\partial t}\right) * \vec{k}$$

Ω = rotation vector of (R_R) with respect to (R_A).

$$\overrightarrow{PM} = X * \vec{I} + Y * \vec{J} + Z * \vec{K}$$

$$\overrightarrow{V_r} = \left(\frac{\partial x}{\partial t}\right) * \vec{I} + \left(\frac{\partial y}{\partial t}\right) * \vec{J} + \left(\frac{\partial z}{\partial t}\right) * \vec{K} \text{ (velocity of point M in } (R_R))$$

The velocity of point M in ($V_r = 0$) would have a speed called the drive speed.

$$\overrightarrow{V_e} = \overrightarrow{V_p} + \vec{\Omega} \wedge \overrightarrow{PM}$$

From which, for any point:

$$\overrightarrow{V_a} = \overrightarrow{V_e} + \overrightarrow{V_r}$$

Acceleration

$$\overrightarrow{\Gamma_a} = \overrightarrow{\Gamma_p} + \vec{\Omega} \wedge (\vec{\Omega} \wedge \overrightarrow{PM}) + \left(\frac{\partial \vec{\Omega}}{\partial t}\right) \wedge \overrightarrow{PM} + 2 * \vec{\Omega} \wedge \overrightarrow{V_r} + \overrightarrow{\Gamma_r}$$

with the acceleration of the point P in the absolute frame (R_A).

$$\overrightarrow{\Gamma_p} = \frac{\partial^2 x_p}{\partial t^2} * \vec{\imath} + \left(\frac{\partial^2 y_p}{\partial t^2}\right) * \vec{\jmath} + \left(\frac{\partial^2 z_p}{\partial t^2}\right) * \vec{k}$$

Ω = rotation vector of (R_R) with respect to (R_A).

$$\overrightarrow{PM} = X * \vec{I} + Y * \vec{J} + Z * \vec{K}$$

$$\overrightarrow{V_r} = \left(\frac{\partial x}{\partial t}\right) * \vec{I} + \left(\frac{\partial y}{\partial t}\right) * \vec{J} + \left(\frac{\partial z}{\partial t}\right) * \vec{K}$$

$$\overrightarrow{\Gamma_r} = \frac{\partial^2 x}{\partial t^2} * \vec{I} + \left(\frac{\partial^2 y}{\partial t^2}\right) * \vec{J} + \left(\frac{\partial^2 z}{\partial t^2}\right) * \vec{K}$$

is the acceleration of point M in the reference (R_R).

NOTE.– $\overrightarrow{V_r}$ and $\overrightarrow{\Gamma_r}$ are the speed and the acceleration of the point M, identified in (R_R), respectively.

The order in which the terms of Γ are written is not random. It is customary to group the first three together under the designation of drive acceleration Γ_e (see the definition of drive movement in section 5.3.3). In fact, if the point M is immobile in the reference (R_R) so that $\overrightarrow{\Gamma_r} = 0$, then

$$\overrightarrow{Vr} = \vec{0}$$

and:

$$\overrightarrow{\Gamma_r} = \vec{0}$$

Therefore, by definition:

$$\overrightarrow{\Gamma_e} = \overrightarrow{\Gamma_p} + \vec{\Omega} \wedge (\vec{\Omega} \wedge \overrightarrow{PM}) + \left(\frac{\partial x}{\partial t}\right) \wedge \overrightarrow{PM}$$

which is the so-called drive acceleration.

The absolute acceleration is finally written as:

$$\overrightarrow{\Gamma_a} = \overrightarrow{\Gamma_e} + 2 * \vec{\Omega} \wedge \overrightarrow{Vr} + \overrightarrow{\Gamma r}$$

where $2 * \vec{\Omega} \wedge \vec{Vr}$ represents the coupling of the rotation of (R_R with respect to (R_A) and velocity in (R_R)). This term is called Coriolis acceleration. It is involved, for example, in the movement of clouds around depressions (Coriolis 1835).

5.4. Components of the instantaneous velocity vector of the aircraft reference (O, i_e, j_e, k_e)$_E$ with respect to the local geographic reference (O, i_{lgt}, j_{lgt}, k_{lgt})$_{LGT}$ in the aircraft reference

(I, U, J, V) in the same plane ⊥ to (O, K).

(U, I_1, K, K') in the same plane ⊥ to (O, V).

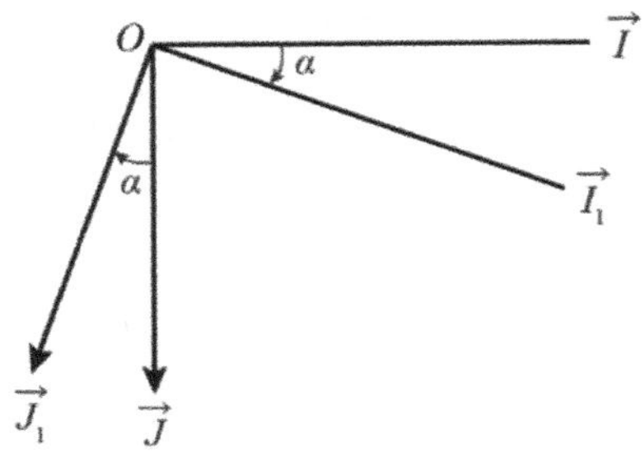

Figure 5.4. *Instantaneous velocity vector components*

COMMENTS ON FIGURE 5.4.–

$$\vec{I} = \cos\alpha * \vec{I_1} - \sin\alpha * \vec{J_1}$$

$$\vec{J} = \sin\alpha * \vec{I_1} + \cos\alpha * \vec{J_1}$$

$$\vec{I1} = \cos\alpha * \vec{I} + \sin\alpha * \vec{J}$$

$$\vec{J1} = -\sin\alpha * \vec{I} + \cos\alpha * \vec{J}$$

$(O, \vec{i}, \vec{j}, \vec{k})$ = *local Galilean trihedron*

$(O, \vec{i_e}, \vec{j_e}, \vec{k_e})$ = *aircraft trihedron*

We can go from the Galilean frame of reference to the aircraft trihedron with three successive rotations:

– $(O, \vec{\imath}, \vec{\jmath}, \vec{k})_{Gal} \rightarrow (O, \vec{u}, \vec{v}, \vec{k})_{R1}$ by rotation ψ around $(O, \vec{k})$;

– $(O, \vec{u}, \vec{v}, \vec{k})_{R1} \rightarrow (O, \vec{\imath_e}, \vec{v}, \vec{k'})_{R2}$ by rotation θ around $(O, \vec{v})$;

– $(O, \vec{\imath_e}, \vec{v}, \vec{k'})_{R2} \rightarrow (O, \vec{\imath_e}, \vec{v}, \vec{k'})_E$ by rotation φ around $(O, \vec{\imath_e})$.

The speed vector of a point on the aircraft is zero with respect to the aircraft trihedron. With respect to the trihedron $(O, \vec{\imath_e}, \vec{v}, \vec{k'})_E$, this speed vector is written:

$$\vec{V}_{R2}(M) = \left(\frac{\partial \phi}{\partial t}\right) * \vec{\imath_e} \wedge \overrightarrow{OM}$$

The speed vector of a point on the aircraft is zero with respect to the aircraft trihedron and, with respect to the trihedron $(O, \vec{u}, \vec{v}, \vec{k})_{R1}$, it is necessary to add the drive speed of a reference $(O, \vec{\imath_e}, \vec{v}, \vec{k'})_{R2}$. With respect to the reference $(O, \vec{u}, \vec{v}, \vec{k})_{R1}$, this speed vector is written as:

$$\vec{V}_{R1}(M) = \left(\frac{\partial \phi}{\partial t}\right) * \vec{\imath_e} \wedge \overrightarrow{OM} + \left(\frac{\partial \theta}{\partial t}\right) * \vec{v} \wedge \overrightarrow{OM}$$

The speed vector of a point of the aircraft with respect to the Galilean trihedron is written by adding the drive speed of a point of the reference point $(O, \vec{u}, \vec{v}, \vec{k})_{R1}$. With respect to the reference $(O, \vec{\imath}, \vec{\jmath}, \vec{k})_{Gal}$:

$$\vec{V}_{Gal}(M) = \left(\frac{\partial \phi}{\partial t}\right) * \vec{\imath_e} \wedge \overrightarrow{OM} + \left(\frac{\partial \theta}{\partial t}\right) * \vec{v} \wedge \overrightarrow{OM} + \left(\frac{\partial \Psi}{\partial t}\right) * \vec{k} \wedge \overrightarrow{OM}$$

We can write:

$$\vec{V}_{Gal}(M) = \vec{\Omega}_{Gal} \wedge \overrightarrow{OM}$$

with:

$$\vec{\Omega}_{Gal} = \left(\frac{\partial \phi}{\partial t}\right) * \vec{\imath_e} + \left(\frac{\partial \theta}{\partial t}\right) * \vec{v} + \left(\frac{\partial \Psi}{\partial t}\right) * \vec{k}$$

The vector $\vec{\Omega}_{Gal}$ is the aircraft rotation speed vector in its movement with respect to the Galilean frame of reference $(O,\vec{i},\ \vec{j},\ \vec{k})_{Gal}$.

The angle rotation Ψ around k, in the base (u,v,k), is given by the matrix:

$$\begin{pmatrix} i \\ j \\ k \end{pmatrix} = \begin{pmatrix} \cos\Psi & -\sin\Psi & 0 \\ \sin\Psi & \cos\Psi & 0 \\ 0 & 0 & 1 \end{pmatrix} \begin{pmatrix} u \\ v \\ k \end{pmatrix}$$

with:

$$\vec{u} \in [\vec{i},\vec{j}],\ \vec{i} \in [\vec{u},-\vec{v}],\ \vec{v} \in [-\vec{i},\vec{j}] \text{ and } \vec{j} \in [\vec{u},\vec{v}]$$

The angle rotation θ around v in the base $(O,\vec{i_e},\vec{v},\vec{k'})_{R2}$ is given by the matrix:

$$\begin{pmatrix} u \\ v \\ k \end{pmatrix} = \begin{pmatrix} \cos\theta & 0 & \sin\theta \\ 0 & 1 & 0 \\ -\sin\theta & 0 & \cos\theta \end{pmatrix} \begin{pmatrix} i_e \\ v \\ k' \end{pmatrix}$$

$$\vec{u} \in [\vec{i_e},\vec{k'}] \text{ and } \vec{k} \in [-\vec{i_e},\vec{k'}]$$

The angle rotation around i_e in the base $(O,\vec{i_e},\vec{v},\vec{k'})_{R2}$ is given by the matrix:

$$\begin{pmatrix} i_e \\ v \\ k' \end{pmatrix} = \begin{pmatrix} 1 & 0 & 0 \\ 0 & \cos\phi & -\sin\phi \\ 0 & \sin\phi & \cos\phi \end{pmatrix} \begin{pmatrix} i_e \\ j_e \\ k_e \end{pmatrix}$$

$$\vec{v} \in [\vec{j_e},-\vec{k_e}] \text{ and } \vec{k'} \in [\vec{j_e},\vec{k_e}]$$

We obtain:

$$\vec{v} = \cos\phi * \vec{j_e} - \sin\phi * \vec{k_e}$$

$$\vec{k'} = \sin\phi * \vec{j_e} + \cos\phi * \vec{k_e}$$

$$\vec{k} = -\sin\theta * \vec{i_e} + \cos\theta * \vec{k'}$$
$$= -\sin\theta * \vec{i_e} + \cos\theta * (\sin\phi * \vec{j_e} + \cos\phi * \vec{k_e})$$

with:

$$\vec{V}_{Gal}(M) = \left(\frac{\partial\phi}{\partial t}\right) * \vec{\imath_e} \wedge \overrightarrow{OM} + \left(\frac{\partial\theta}{\partial t}\right) * \vec{v} \wedge \overrightarrow{OM} + \left(\frac{\partial\psi}{\partial t}\right) * \vec{k} \wedge \overrightarrow{OM}$$

$$\vec{V}_{Gal}(M) = \left(\frac{\partial\phi}{\partial t}\right) * \vec{\imath_e} \wedge \overrightarrow{OM} + \left(\frac{\partial\theta}{\partial t}\right) * (\cos\phi * \vec{\jmath_e} - \sin\phi * \vec{k_e}) \wedge \overrightarrow{OM} + \left(\frac{\partial\psi}{\partial t}\right) * (-\sin\theta * \vec{\imath_e} + \cos\theta * (\sin\phi * \vec{\jmath_e} + \cos\phi * \vec{k_e})) \wedge \overrightarrow{OM}$$

The components p_e, q_e and r_e of the instantaneous speed vector $\vec{\Omega}_{E/Lgt}$ of rotation of the aircraft reference in the base $(O, i_e, j_e, k_e)_E$, in relation to the local geographic reference, are written in the aircraft reference in the form:

$$p_e = \left(\frac{\partial\phi}{\partial t}\right) - \left(\frac{\partial\Psi}{\partial t}\right) * \sin\theta$$

$$q_e = \left(\frac{\partial\theta}{\partial t}\right) * \cos\phi + \left(\frac{\partial\psi}{\partial t}\right) * \cos\theta * \sin\phi$$

$$r_e = \left(\frac{\partial\psi}{\partial t}\right) * \cos\theta * \cos\phi - \left(\frac{\partial\theta}{\partial t}\right) * \sin\phi$$

The components p'_e, q'_e and r'_e of the instantaneous acceleration vector $\vec{\Gamma}_{Gal}$ in the aircraft reference $(O, \vec{\imath}_e, \vec{\jmath}_e, \vec{k}_e)_E$ are given by differentiating with respect to time the equations p_e, q_e and r_e of the instantaneous velocity vector (see section 5.4):

$$p_e = \left(\frac{\partial\phi}{\partial t}\right) - \left(\frac{\partial\Psi}{\partial t}\right) * \sin\theta$$

$$q_e = \left(\frac{\partial\theta}{\partial t}\right) * \cos\phi + \left(\frac{\partial\psi}{\partial t}\right) * \cos\theta * \sin\phi$$

$$r_e = \left(\frac{\partial\psi}{\partial t}\right) * \cos\theta * \cos\phi - \left(\frac{\partial\theta}{\partial t}\right) * \sin\phi$$

$$p'_e = \left(\frac{\partial p_e}{\partial t}\right) = -\cos\theta * \left(\frac{\partial\theta}{\partial t}\right) * \left(\frac{\partial\psi}{\partial t}\right) - \sin\theta * \left(\frac{\partial^2\psi}{\partial t^2}\right) + \left(\frac{\partial^2\phi}{\partial t^2}\right)$$

$$q'_e = \left(\frac{\partial q_e}{\partial t}\right) = -\sin\theta * \left(\frac{\partial\theta}{\partial t}\right) * \sin\phi * \left(\frac{\partial\psi}{\partial t}\right) + \cos\theta * \cos\phi * \left(\frac{\partial\phi}{\partial t}\right) * \left(\frac{\partial\psi}{\partial t}\right) + \cos\theta * \sin\phi * \left(\frac{\partial^2\psi}{\partial t^2}\right) - \sin\phi * \left(\frac{\partial\phi}{\partial t}\right) * \left(\frac{\partial\theta}{\partial t}\right) + \cos\phi * \left(\frac{\partial^2\theta}{\partial t^2}\right)$$

$$r'_e = \left(\frac{\partial re}{\partial t}\right) = -\sin\theta * \left(\frac{\partial\theta}{\partial t}\right) * \cos\phi * \left(\frac{\partial\psi}{\partial t}\right) - \cos\theta * \sin\phi * \left(\frac{\partial\phi}{\partial t}\right) * \left(\frac{\partial\psi}{\partial t}\right) + \cos\theta * \cos\phi * \left(\frac{\partial^2\psi}{\partial t^2}\right) - \cos\phi * \left(\frac{\partial\phi}{\partial t}\right) * \left(\frac{\partial\theta}{\partial t}\right) - \sin\phi * \left(\frac{\partial^2\theta}{\partial t^2}\right)$$

5.5. Equations of accelerations and forces in the aircraft frame $(O, i_e, j_e, k_e)_E$

Let $(O, \vec{i}_e, \vec{j}_e, \vec{k}_e)_E$ be the reference linked to the aircraft. The forces of inertia are defined by the vector derived from the momentum. In the aircraft reference, they are by definition:

$$(\overrightarrow{-F_{inertia}})_E = \left(\frac{\partial m * \vec{V}}{\partial t}\right)_{Gal} = m * \left(\frac{\partial \vec{V}}{\partial t}\right)_{Gal} = m * \left(\frac{\partial \vec{V}}{\partial t}\right)_E + m * \vec{\Omega}_{E/Gal} \wedge \vec{V}_E$$

with G_{al} being the original Galilean reference mark of the test track; for example, with the aircraft defined as follows:

– the OX axis is directed east;

– the OY axis is directed north;

– the OZ axis, which is the descending vertical, completes this direct trihedron.

The aircraft reference is defined as follows:

– the OX axis is directed along the longitudinal axis of the aircraft (positive forward);

– the OY axis is directed perpendicular to the OX axis and positive to the right;

– the OZ axis completes this direct trihedron.

The point of origin of the aircraft reference point is the forward point of the aircraft, the center of gravity entered negatively at X.

$\vec{\Omega}_{E/Gal}$ is the speed vector of the aircraft at the center of gravity.

By expressing the derivative of the speed vector $\vec{V}$ in the trihedron of the aircraft $(O, \vec{i}_e, \vec{j}_e, \vec{k}_e)_E$, we can express the forces of inertia in this aircraft reference frame in the form of the following equations:

$$-(X_{inertia})_E = m * \left(\left(\frac{\partial u_e}{\partial t}\right) + q_e * w_e - r_e * v_e\right)$$

$$-(Y_{inertia})_E = m * \left(\left(\frac{\partial v_e}{\partial t}\right) + r_e * u_e - p_e * w_e\right)$$

$$-(Z_{inertia})_E = m * \left(\left(\frac{\partial w_e}{\partial t}\right) + p_e * v_e - q_e * u_e\right)$$

In the aircraft trihedron $(O, \vec{i}_e, \vec{j}_e, \vec{k}_e)_E$, we can express the accelerations as follows:

$$-(\Gamma_{ie})_E = m * \left(\left(\frac{\partial u_e}{\partial t}\right) + q_e * w_e - r_e * v_e\right)$$

$$-(\Gamma_{je})_E = m * \left(\left(\frac{\partial v_e}{\partial t}\right) + r_e * u_e - p_e * w_e\right)$$

$$-(\Gamma_{ke})_E = m * \left(\left(\frac{\partial w_e}{\partial t}\right) + p_e * v_e - q_e * u_e\right)$$

The general equations in the base $(O, \vec{i}_e, \vec{j}_e, \vec{k}_e)_E$ give:

$$-(X_{inertia})_E = (R_x)_E + (T_x)_E + m * g * \sin\theta$$

$$-(Y_{inertia})_E = (R_y)_E + (T_y)_E - m * g * \cos\theta * \sin\phi$$

$$-(Z_{inertia})_E = (R_z)_E + (T_z)_E - m * g * \cos\theta * \cos\phi$$

with:

– $(R_x)_E = (R_x)_{Gal}$, the drag force of the aircraft;

– $(R_y)_E = (R_y)_{Gal}$, the lateral lift force of the aircraft;

– $(R_z)_E = (R_z)_{Gal}$, the lift force of the aircraft ($R_z > 0$ if $Z < 0$);

– $(T_x)_E = (PHI_x)_{Gal} + F_x$, the thrust force of the aircraft in the OX axis of the aircraft reference;

– $(T_y)_E = (PHI_y)_{Gal} + F_y$, the thrust force of the aircraft in the OY axis of the aircraft reference;

– $(T_z)_E = (PHI_z)_{Gal}$, the thrust force of the aircraft in the OZ axis of the aircraft reference.

Furthermore:

– F designates the engine thrust assumed to be centered in the aircraft yaw plane (FZ = 0);

– PHI is the thrust of accelerators.

From these relationships, we can express the linear accelerations in the aircraft reference as:

$$\left(\frac{\partial u_e}{\partial t}\right) = -g * \sin\theta - q_e * w_e + r_e * v_e + \frac{PHI_x + F_x + R_x}{m}$$

$$\left(\frac{\partial v_e}{\partial t}\right) = g * \cos\theta * \sin\phi - r_e * u_e + p_e * w_e + \frac{PHI_y + F_y + R_y}{m}$$

$$\left(\frac{\partial w_e}{\partial t}\right) = g * \cos\theta * \cos\phi - p_e * v_e + q_e * u_e + \frac{PHI_z + R_z}{m}$$

5.6. Determination of load factors in the aircraft reference $(O, i_e, j_e, k_e)_E$

The load factor is defined as the ratio of the resultant of forces external to the aircraft (except the weight of the aircraft) and the mass m of the aircraft at time t.

In the aircraft trihedron, with $(O, \vec{i}_e, \vec{j}_e, \vec{k}_e)_E$, we can express the load factors along the three axes as follows:

$$(n_x)_E = \frac{1}{g}\left((g * \sin\theta) - (r_e * v_e - q_e * w_e) + \left(\frac{\partial u_e}{\partial t}\right)\right) = \frac{PHI_x + F_x + R_x}{m * g}$$

$$(n_y)_E = \frac{1}{g} * \left((-g) * \cos\theta * \sin\phi - (p_e * w_e - r_e * u_e) + \left(\frac{\partial v_e}{\partial t}\right)\right) + \frac{PHI_y + F_y + R_y}{m * g}$$

$$(n_z)_E = \frac{1}{g} * \left((-g) * \cos\theta * \cos\phi - (q_e * u_e - p_e * v_e) + \left(\frac{\partial w_e}{\partial t}\right)\right) + \frac{PHI_z + R_z}{m * g}$$

5.7. Case where the speeds and accelerations are provided at a point other than G

According to section 5.3.6, if the speed is known at the level of the inertial unit of the aircraft, then the speed at the center of gravity is given by the following relation:

$$\vec{V_e} = \vec{V_p} + \vec{\Omega} \wedge \overrightarrow{PM}$$

The center of gravity being the fixed point of the aircraft reference, then:

$$\vec{V_e} = \vec{V_p} + \vec{\Omega} \wedge \overrightarrow{PM}$$

becomes:

$$\overrightarrow{V_G} = \overrightarrow{V_C} + \vec{\Omega} \wedge \overrightarrow{CG}$$

or:

$$\overrightarrow{V_G} = \overrightarrow{V_C} - \vec{\Omega} \wedge \overrightarrow{GC}$$

The speeds of the center of gravity in the aircraft reference frame are given by the following relations:

$$(U_G)_E = (U_C)_E - Q_e * (Z_C - Z_G) + R_e * (Y_C - Y_G)$$

$$(V_G)_E = (V_C)_E - R_e * (X_C - X_G) + P_e * (Z_C - Z_G)$$

$$(W_G)_E = (W_C)_E - P_e * (Y_C - Y_G) + Q_e * (X_C - X_G)$$

Likewise, drive acceleration is written as:

$$\overrightarrow{\Gamma_e} = \overrightarrow{\Gamma_p} + \vec{\Omega} \wedge (\vec{\Omega} \wedge \overrightarrow{PM}) + \left(\frac{\partial \vec{\Omega}}{\partial t}\right) \wedge \overrightarrow{PM}$$

The accelerations of the center of gravity in the aircraft frame are given by the following relations:

$$\left(\frac{\partial (U_G)_E}{\partial t}\right) = \left(\frac{\partial (U_C)_E}{\partial t}\right) - \dot{Q}_e * (Z_C - Z_G) + \dot{R}_e * (Y_C - Y_G)$$
$$+(Q_e^2 + R_e^2) * (X_C - X_G) - P_e * Q_e * (Y_C - Y_G) - P_e * R_e * (Z_C - Z_G)$$

$$\left(\frac{\partial (V_G)_E}{\partial t}\right) = \left(\frac{\partial (V_C)_E}{\partial t}\right) - \dot{R}_e * (X_C - X_G) + \dot{P}_e * (Z_C - Z_G)$$
$$+(P_e^2 + R_e^2) * (Y_C - Y_G) - R_e * Q_e * (Z_C - Z_G) - P_e * Q_e * (X_C - X_G)$$

$$\left(\frac{\partial (W_G)_E}{\partial t}\right) = \left(\frac{\partial (W_C)_E}{\partial t}\right) - \dot{P}_e * (Y_C - Y_G) + \dot{Q}_e * (X_C - X_G)$$
$$+(P_e^2 + Q_e^2) * (Z_C - Z_G) - P_e * R_e * (X_C - X_G) - Q_e * R_e * (Y_C - Y_G)$$

The load factors are also given at the center of gravity by the following relations:

$$(n_x)_E = \frac{1}{g} * \left(\left(\frac{\partial (U_G)_E}{\partial t} \right) - (g * (-\sin \theta)) \right) = \frac{PHIX + Fx + Rx}{m * g}$$

$$(n_y)_E = \frac{1}{g} * \left(\left(\frac{\partial (V_G)_E}{\partial t} \right) - (g * \cos \theta * \sin \phi) \right) = \frac{PHIY + Fy + Ry}{m * g}$$

$$(n_z)_E = \frac{1}{g} * \left(\left(\frac{\partial (W_G)_E}{\partial t} \right) - (g * \cos \theta * \cos \phi) \right) = \frac{PHIZ + Rz}{m * g}$$

5.8. Coordinates of the aerodynamic speed in aircraft axes

Let $\vec{V}$ be the aerodynamic speed vector of the center of gravity of the aircraft and (u_e, v_e, w_e) its projections on the axes of the aircraft (G, x_e, y_e, z_e).

In the aerodynamic frame, the projections on the axes of the aerodynamic speed vector $\vec{V}_{ae}(G, x_{ae}, y_{ae}, z_{ae})$ are $(V_{ae}, 0, 0)$:

$$\begin{pmatrix} u_e \\ v_e \\ w_e \end{pmatrix} = \begin{pmatrix} V_{ae} * \cos j * \cos i \\ V_{ae} * \sin j \\ V_{ae} * \cos j * \sin i \end{pmatrix}$$

The instantaneous rotation speed vector of the aircraft relative to the fixed axes is called $\vec{\Omega}$.

The projections on the aircraft axes (G, x e, y_e, z_e) are p_e, q_e, r_e, on the aerodynamic velocity axes (G, x_e, y_e, z_e) they are p_e, q_e, r_e and on the aerodynamic velocity axes $(G, x_{ae}, y_{ae}, z_{ae})$ they are p_{ae}, q_{ae}, r_{ae}.

$$\begin{pmatrix} p_{ae} \\ q_{ae} \\ r_{ae} \end{pmatrix} = \begin{pmatrix} p_e * \cos j * \cos i + q_e * \sin j + r_e * \cos j * \sin i \\ -p_e * \sin j * \cos i + q_e * \cos j - r_e * \sin j * \sin i \\ -p_e * \sin i + r_e * \cos i \end{pmatrix}$$

$$\begin{pmatrix} p_e \\ q_e \\ r_e \end{pmatrix} = \begin{pmatrix} p_{ae} * \cos j * \cos i - q_{ae} * \sin j * \cos i - r_{ae} * \sin i \\ p_{ae} * \sin j + q_{ae} * \cos j \\ p_{ae} * \cos j * \sin i - q_{ae} * \sin j * \sin i + r_{ae} * \cos i \end{pmatrix}$$

The projections (p_e, q_e, r_e) on the aircraft axes (G, x_e, y_e, z_e) are given in section 5.3.6.

5.9. Equations of moments in the aircraft frame (O, i$_e$, j$_e$, k$_e$)$_E$

The mass m alone does not allow us to characterize the difficulty in putting a solid in motion. In the particular case where this movement is a translation, the mass is sufficient, but for more complex movements, the distribution of this mass on the solid must be maintained.

Two scalar quantities, the moment of inertia and the product of inertia, characterize this distribution around an axis. These two quantities are expressed in kg.m^2, and we detail their properties below.

The inertia couples are defined by the vector derived from the angular momentum.

In the aircraft trihedron $(O, \vec{i}_e, \vec{j}_e, \vec{k}_e)_E$, we can determine the angular momentum using the barycentric frame of reference $(O, \vec{i}_{bary}, \vec{j}_{bary}, \vec{k}_{bary})_{bary}$ and Koenig's two theorems.

The angular momentum at O (center of the Galilean frame) is given by:

$$\vec{K}_O = \vec{K}_{bary\,G} + \overrightarrow{OG} \wedge \vec{p}$$

with:

$$\vec{p} = m * \vec{V}$$

$\vec{K}_{2G}$ is the angular momentum in G for the motion in the barycentric frame of reference. In this frame of reference, the solid has a rotational movement around the mass center of gravity or G.

If $\overrightarrow{\Omega}$ is the instantaneous speed of the rotation vector, the speed of a point P of the solid is given by:

$$\vec{V}(P) = \overrightarrow{PG} \wedge \vec{\Omega}$$

The angular momentum in G is:

$$\vec{K}_{bary\ G} = \iiint_{(S)} \left(\overrightarrow{GP} \wedge \overrightarrow{V(P)}\right) dm = \iiint_{(S)} \left(\overrightarrow{GP} \wedge (\overrightarrow{PG} \wedge \vec{\Omega}\right) dm$$

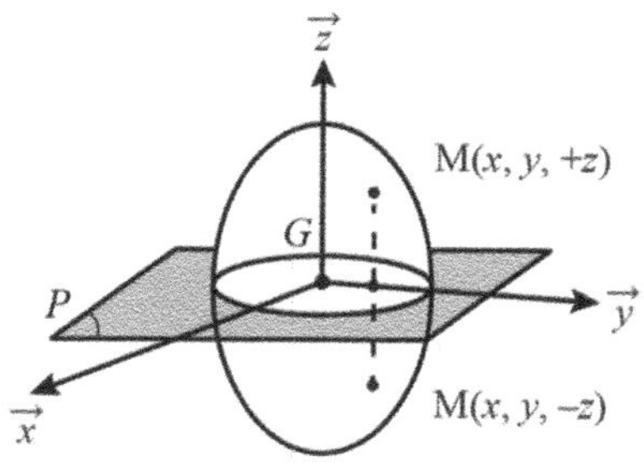

Figure 5.5. *Solid admitting planes of symmetry*

In the barycentric basis $(O, \vec{i}_{bary}, \vec{j}_{bary}, \vec{k}_{bary})_{bary}$, this relation is expressed in the matrix form:

$$\begin{pmatrix} K_{bary\ GX} \\ K_{bary\ GY} \\ K_{bary\ GZ} \end{pmatrix} = \begin{pmatrix} J_{xx} & J_{xy} & J_{xz} \\ J_{yx} & J_{yy} & J_{yz} \\ J_{zx} & J_{zy} & J_{zz} \end{pmatrix} \begin{pmatrix} \Omega_x \\ \Omega_y \\ \Omega_z \end{pmatrix}$$

with:

– J_{xx}: moment of inertia of the solid with respect to $(O, \overrightarrow{I_E})$, defined by:

$$\int_{(S)} ((y^2 + z^2) * dm)$$

– J_{yy}: moment of inertia of the solid with respect to $(O, \overrightarrow{J_E})$, defined by:

$$\int_{(S)} ((x^2 + z^2) * dm)$$

– J_{zz}: moment of inertia of the solid with respect to $(O, \overrightarrow{z_E})$, defined by:

$$\int_{(S)} ((x^2 + y^2) * dm)$$

– $J_{xy} = J_{yx}$: product of inertia with respect to the plane $(O, \overrightarrow{\imath_E}, \overrightarrow{\jmath_E})$, defined by:

$$\int_{(S)} ((x * y * dm)$$

– $J_{xz} = J_{zx}$: product of inertia with respect to the plane $(O, \overrightarrow{\imath_E}, \overrightarrow{z_E})$, defined by:

$$\int_{(S)} ((x * z * dm)$$

– $J_{yz} = J_{zy}$: product of inertia with respect to the plane $(O, \overrightarrow{\jmath_E}, \overrightarrow{z_E})$, defined by:

$$\int_{(S)} ((y * z * dm)$$

In the case of a cruciform aircraft, $(O, \overrightarrow{\imath_E}, \overrightarrow{\jmath_E})$ is the material plane of symmetry of the normal $\overrightarrow{\jmath_E}$ for the solid.

We can then separate the integral over (S) into a sum of integrals over (S1) and (S2):

$$\int_{(S1)} (z * x * dm) + \int_{(S2)} ((-z) * x * dm) = 0$$

$$\int_{(S1)} (y * z * dm) + \int_{(S2)} ((-y) * z * dm) = 0$$

In the case of a cruciform aircraft, $(O, \overrightarrow{\imath_E}, \overrightarrow{\jmath_E})$ is the material plane of symmetry of the normal z_e for the solid.

We can then separate the integral over (S) into a sum of integrals over (S1) and (S2):

$$\int_{(S1)} (y * z * dm) + \int_{(S2)} ((-y) * z * dm) = 0$$

$$\int_{(S1)} (y * z * dm) + \int_{(S2)} ((-y) * z * dm) = 0$$

or

$$\vec{K}_{bary\ GX} = K_{bary\ GX} * \vec{\Omega}.$$

The expression of the inertia matrix is only simple if we choose the main inertial axes linked to the solid.

In the aircraft trihedron $(O, \vec{i_e}, \vec{J_e}, \vec{z_e})_E$, $\vec{W} = p_e * \vec{i_e} + q_e * \vec{j_e} + r_e * \vec{k_e}$

For the cruciform aircraft, we have:

$$J_{bary\ G} = \begin{pmatrix} J_{xx} & 0 & 0 \\ 0 & J_{yy} & 0 \\ 0 & 0 & J_{zz} \end{pmatrix}$$

and:

$$\vec{K}_{bary\ G} = J_{bary\ G} * \vec{\Omega}$$

or:

$$\vec{K}_{bary\ G} = J_{xx} * p_e * \vec{i_e} + J_{yy} * q_e * \vec{j_e} + J_{yy} * r_e * \vec{k_e}$$

For the classic aircraft (airplane type), we have:

$$J_{bary\ G} = \begin{pmatrix} J_{xx} & 0 & J_{xz} \\ 0 & J_{yy} & 0 \\ J_{zx} & 0 & J_{zz} \end{pmatrix}$$

and:

$$\vec{K}_{bary\ G} = J_{bary\ G} * \vec{\Omega}$$

or:

$$\overrightarrow{K}_{bary\,G} = (J_{xx} * p_e - J_{zx} * r_e) * \overrightarrow{i_e} + J_{yy} * q_e * \overrightarrow{j_e} + ((-J_{zx}) * p_e + J_{zz} * r_e) * \overrightarrow{k_e}$$

The derivative of angular momentum is expressed as follows:

$$\left(\frac{\partial \overrightarrow{K}_{bary}}{\partial t}\right)_{bary} = \left(\frac{\partial \overrightarrow{K}_{bary}}{\partial t}\right)_{bary} + \Omega_{Kbary/E} {}^\wedge \overrightarrow{K}_{bary}$$

$$\left(\frac{\partial \overrightarrow{K}_{bary}}{\partial t}\right)_{bary} = \left(J_{xx} * \frac{\partial p_e}{\partial t} - J_{zx} * \frac{\partial r_e}{\partial t}\right)_E * \overrightarrow{i_e} + \left(J_{yy} * \frac{\partial q_e}{\partial t}\right)_E * \overrightarrow{j_e} + \left((-J_{zx}) * \frac{\partial p_e}{\partial t} + J_{zz} * \frac{\partial r_e}{\partial t}\right)_E * \overrightarrow{k_e} + \left(J_{zz} * r_e * q_e - J_{zx} * p_e * q_e - J_{yy} * r_e * q_e\right)_E * \overrightarrow{i_e} + (J_{zx} * (p_e^2 - r_e^2) + (J_{xx} - J_{zz}) * r_e * p_e)_E * \overrightarrow{j_e} + \left(J_{yy} - J_{xx}) * p_e * q_e + J_{zx} * q_e * r_e\right)_E * \overrightarrow{k_e}$$

This relation expresses the fact that the vector derived from the angular momentum in G in the barycentric frame of reference (bary) is written in the base of the aircraft frame of reference by the above equation.

The moments of inertia expressed in the aircraft frame are therefore, taking $(G, \overrightarrow{i_e})$, $(G, \overrightarrow{j_e})$ and $(G, \overrightarrow{k_e})$ as the main axes of inertia:

$$-L_{Einertia} = J_{xx} * \frac{\partial p_e}{\partial t} + (J_{zz} - J_{yy}) * r_e * q_e$$

$$-M_{Einertia} = J_{yy} * \frac{\partial q_e}{\partial t} + (J_{xx} - J_{zz}) * r_e * p_e$$

$$-N_{Einertia} = J_{zz} * \frac{\partial r_e}{\partial t} + (J_{yy} - J_{xx}) * p_e * q_e$$

The general equations of the balance in the aircraft base $(G, \overrightarrow{i_e}, \overrightarrow{j_e}, \overrightarrow{k_e})$ give:

$$-L_{Einertia} = L_E + L_{ET}$$

$$-M_{Einertia} = M_E + M_{ET}$$

$$-N_{Einertia} = N_E + N_{ET}$$

with:

$$L_E = G_{MX}, M_E = G_{My} \text{ and } N_E = G_{MZ} \text{ (aerodynamic torques)}$$

and:

$$L_{ET} = CPHIX - F_Y * (Z_F - Z_G) \text{ (torques due to propulsion)}$$

$$M_{ET} = CPHIY - F_X * (Z_F - Z_G)$$

$$L_{ET} = CPHIX$$

From these relations, we obtain:

$$\frac{\partial p_e}{\partial t} = \frac{1}{J_{xx}} * (-(J_{zz} - J_{yy}) * r_e * q_e + G_{MX} + CPHIX - F_Y * (Z_F - Z_G)$$

$$\frac{\partial q_e}{\partial t} = \frac{1}{J_{yy}} * (-(J_{xx} - J_{zz}) * r_e * p_e + G_{MY} + CPHIY - F_X * (Z_F - Z_G)$$

$$\frac{\partial r_e}{\partial t} = \frac{1}{J_{zz}} * (-(J_{yy} - J_{xx}) * p_e * q_e + G_{MZ} + CPHIZ$$

5.10. Forces and moments applied to the aircraft

5.10.1. *Force of gravity*

The force of gravity is a downward vertical force of intensity $m * g$, with m being the aircraft mass.

The weight of the aircraft in the aircraft reference is given by the following equations:

$$-(m * g * \sin\theta) \text{ projection on vector } \vec{i_e}$$

$$+(m * g * \cos\theta * \sin\phi) \text{ projection on vector } \vec{j_e}$$

$$+(m * g * \cos\theta * \cos\phi) \text{ projection on vector } \vec{k_e}$$

In the aircraft trihedron $(G, \vec{i_e}, \vec{j_e}, \vec{k_e})_E$, we can express the load factors along the three axes of the aircraft.

5.10.2. *Forces and moments of propulsion*

These are the tensile forces exerted by the thruster (or thrusters) and the associated moments:

– on G_{XeYeZe}: T_{Xe}, T_{Ye}, T_{Ze} (see section 6.6) and L_{eT}, M_{eT}, N_{eT} (see section 5.9);

– on $G_{XaeYaeZae}$: $T_{Xae}, T_{Yae}, T_{Zae}$(see section 6.6) and $L_{aeT}, M_{aeT}, N_{aeT}$ (see section 5.9).

5.10.3. *Aerodynamic forces and moments*

These are the aerodynamic forces and moments exerted on the aircraft:

– on G_{XeYeZe}: R_{Xe}, R_{Ye}, R_{Ze} (see section 6.6) and L_e, M_e, N_e (see section 5.9);

– on $G_{XaeYaeZae}$: $R_{Xae}, R_{Yae}, R_{Zae}$ and L_{ae}, M_{ae}, N_{ae}.

5.10.4. *Forces and inertia torques*

The forces and moments of inertia have been defined in the aircraft marker G_{XeYeZe}.

The forces (see section 5.5) and moments of inertia (see section 5.9) are defined by the formulas of classical rational mechanics (vector derived from the momentum and the angular momentum).

6

Aircraft Characteristics

6.1. Aerodynamics problems posed by aircraft

The fields of application of aircraft are varied. No single aircraft can cover them all, so there is no "ideal" configuration with perfect characteristics to meet all needs.

The problems to be solved for the design of a civilian aircraft, a military aircraft, a microlight or a drone are not of the same nature. Depending on the needs, an optimum for one or more characteristics will be sought to the detriment of others, considered secondary.

Commonly, the search for optimum qualities concerns the following points:

– the drag;

– maneuverability or the creation of lift;

– stability;

– the resistance of structures from mechanical and thermal points of view.

6.1.1. *Drag*

In order to have long range or achieve maximum speeds, an aircraft must have as little drag as possible.

Most of a flight (the so-called "cruising" flight) is carried out at a load factor n = 1 and, therefore, at low incidence, without significant changes other than path corrections. In this case, the search for a minimum drag coefficient is a priority. This drag is divided into four parts:

– pressure drag (or supersonic wave drag);

– friction drag;

– base drag;

– accessory drag.

The first term indicates that the component, along the longitudinal axis of the integral of the pressures on the aircraft, "base excepted", is not zero. A body drag can be assigned to each element of the aircraft (body, wings, control surfaces, etc.).

The second term is linked to the phenomenon of air viscosity; it varies according to the dimensions of the aircraft, its relative speed and the state of the external surface in contact with the air. This term relates to the viscous flow, which gives rise, over the entire surface of the aircraft, to shear stresses.

The aircraft can be broken down into its various parts, namely:

– the body;

– wings;

– control surfaces.

For each element, the external surfaces in contact with the fluid, called bathed or wet surfaces, must be evaluated. They are related to the reference surface (the maximum diameter of an airplane, for example).

For the body, we have S_{bc}/S_{ref} (=1); for the wings, we have S_{ba}/S_{ref}; and for the rudders, S_{bg}/S_{ref} (with $S_{ref} = \pi * d^2/4$).

By definition, the tangential force T exerted by the fluid on a plate per unit area is given by the equation:

$$T = \mu * \left(\frac{\partial u}{\partial y}\right)_{y=0} = C_f * 0.5 * \rho * V_0^2$$

with:

μ = the fluid viscosity coefficient;

u = the component of the speed according to x in the vicinity of the wall;

y = the axis perpendicular to the wall;

V_0 = the speed of undisturbed flow;

C_f = the friction coefficient.

The third term is due to the appearance of lift-off flow behind the aircraft and the resulting depressions.

A reduction in drag can be obtained by improving, if possible, three components:

– a reduction for the frontal surfaces of each element of the aircraft can be obtained by using a dihedral that is as low as possible for the leading and trailing edges and a very elongated nose cone, which makes it possible to reduce the wave drag of the aircraft;

– the aircraft's friction drag can be improved in very modest proportions by paying attention to the condition of the surfaces in contact with the air;

– it is very difficult to significantly reduce the base drag of the aircraft;

– for on-board accessories (antennas, fairings, etc.), no action is possible to reduce the drag.

Drag reduction is a compromise between the mission of the aircraft, on-board equipment and accessories.

6.1.2. *Lift*

An aircraft must be able to curve its trajectory, which gives it a load factor normal to its speed. This load factor can be produced by an incidence set by aerodynamic means (aerodynamic control surfaces).

Lift appears at the different parts of the aircraft (wings, body, control surfaces, accessories), which has the effect of creating a load factor.

6.1.3. *Reynolds number*

The coefficient of friction C_f is evaluated from the Reynolds number:

$$R_e = \rho * V * \frac{D}{\mu} = V * \frac{D}{\upsilon}$$

with:

V = the flow velocity;

D = the characteristic length (L_{Ref});

υ = the kinematic viscosity of the fluid ($\upsilon = \mu / \rho$);

μ = the dynamic fluid viscosity (Pa·s or kg/(m·s));

ρ = the fluid density (kg/m^3).

The Reynolds number represents the ratio of inertial forces to viscous forces.

The units are given for guidance only, as they are dimensionless numbers with the same value in any coherent system of units.

The Reynolds number also represents the (qualitative) ratio of the transfer by convection to the transfer by diffusion of the momentum.

It is important to note that the Reynolds number only gives an order of magnitude.

The fact that we can test cars (MatraEspace, for example) in a hydrodynamic basin, therefore in a fluid (water) different from their natural fluid (air), shows to what extent the Reynolds number reigns supreme over the flows of all fluids.

6.1.4. *Flow velocity*

The speed of sound in an ideal gas is a function of the Laplace coefficient γ (gamma), of the density ρ (rho) and of the pressure P of the gas and is theoretically calculated as:

$$C_{\text{perfect gas}} = \sqrt{((\gamma * P) / \rho)}$$

with:

$$\gamma = C_P/C_V \text{ Laplace coefficient;}$$

C_P and C_V are the isobaric and isochoric thermal capacities, respectively.

The speed of sound can also be calculated using the specific ideal gas constant:

$$R_S = R / M$$

with M being the molar mass, R the universal constant of ideal gases and T the thermodynamic temperature in Kelvin (K):

$$C_{\text{perfect gas}} = \sqrt{((\gamma * R_S * T)}$$

For air, mainly composed of dioxygen and nitrogen (diatomic gases):

– $R_S = 287$ ($J*kg^{-1}*K^{-1}$);

– $\gamma = 1{,}405$;

– T = temperature at the considered altitude.

For each case, knowing the Mach number, the considered altitude and the corresponding temperature, the speed can be obtained:

$$V = M * \sqrt{((\gamma * R_S * T)} = M * 20.0807 * \sqrt{T}$$

6.1.5. *Stability of an aircraft*

An aircraft is aerodynamically stable when Cmα is negative, in other words, when the focus of aerodynamic forces is behind the center of gravity:

$$-X_{\text{focus}} \left(\frac{C_{m\alpha}}{C_{z\alpha}}\right)_{\alpha=0} * L_{\text{Ref}}$$

This represents the distance between the focus and the center of gravity.

More generally, since aircraft rarely have linear aerodynamic characteristics as a function of the incidence and the roll angle, it is necessary to consider, in order to characterize the stability of an aircraft:

– the position of the application point of the aerodynamic resultant:

$$\frac{(C_m * L_{ref})}{C_Z}$$

– the focus of the aircraft in incidence:

$$X_{f\alpha} = \left(\frac{\partial C_m}{\partial C_z}\right)_\alpha * L_{Ref}$$

– the focus of the aircraft in roll:

$$X_{f\phi} = \left(\frac{\partial C_m}{\partial C_z}\right)_\phi * L_{Ref}$$

The aircraft does not have to be stable throughout its flight envelope; it is the combination of aerodynamic and pilot characteristics that result in either:

– the need to pilot the aircraft with natural and permanent stability;

– the possibility of piloting with artificial stability.

It is through the relative positioning of the bearing surfaces on the body and their dimensioning that the stability of aircraft is regulated.

6.1.6. *Resistance of structures*

Mass gains have immediate repercussions on aircraft performance, such as:

– greater acceleration in the propelled phase, resulting in a speed gain;

– higher load factor for a given lift.

The mechanical elements must therefore be dimensioned as accurately as possible, taking into account the stresses they will have to withstand during all phases of the flight.

A particular point concerns the study of the hinge moments of the control surfaces and the search for solutions to reduce them.

During the flight of an aircraft piloted with the aid of aerodynamic control surfaces, a resulting force $\vec{R}$ and a resulting moment $\vec{M}$ appear on a control surface. These two terms are a function of the Mach number M, the sideslip β, the angle of attack α and the steering angle δ of the rudder.

Low hinge moments allow:

– servomotors with smaller dimensions to be adopted;

– the handling of the aircraft to be improved thanks to the higher turning speeds of the control surfaces (lower response time).

From a thermal point of view, under the effect of the compressibility and the friction of the air, the walls of the aircraft will heat up, hence the need to use materials resistant to temperature.

6.1.7. *Sizing of an aircraft*

Once the load factor requirements of an aircraft are known, it is necessary to size the surfaces, wings and control surfaces, which provide the necessary coefficient of lift C_Z. The load factor requirements are very different depending on the aircraft considered (civil aircraft, military aircraft, surveillance drones, microlights, etc.).

6.2. Load factor

6.2.1. *Definition of the load factor*

By definition, the load factor is the result of dividing the aircraft's acceleration by g, which is 9.81 m/s^2.

$$\text{Load factor} = \frac{\text{acceleration}}{\text{g}}$$

6.2.2. *Definition of the load factor requirements*

For an aircraft that evolves aerodynamically, the process is as follows:

– a deflection of the control surfaces, a deviation of the jet, etc., provide the aircraft with a moment;

– under its action, there is a rotation around the center of gravity and the aircraft is in incidence;

– the effect of the angle of incidence is to create a lift force, $Rz = -(q * S * Cz\alpha * \alpha)$, normal at aircraft speed, which therefore has a load factor:

$$n_z = q * S * C_{z\alpha} * \frac{\alpha}{m * g}$$

Once the load factor requirements of an aircraft are known, it is necessary to dimension the surfaces of the wings and control surfaces, which provide the necessary lift coefficient C_z.

7

Aircraft Simulation Model

We here discuss an observation drone-type aircraft, for which there are several guidance modes.

The first guidance mode is the so-called "remote-controlled" mode, which is provided from the ground. The aircraft then performs the maneuvers commanded by an operator from the ground.

The second mode is the so-called "inertial" mode. In this mode, the control surfaces order maker no longer undertakes the piloting calculations. Guidance and piloting are provided thanks to the information delivered by the drone's inertial unit.

At any time during these inertial phases, the operator has the option of switching to the "remote-controlled" mode.

Indeed, during probationary tests, it may prove useful to ensure the in-flight readjustment of the inertial unit following the observation of discrepancies between the simulated flight and the actual flight. This adjustment is carried out by flying over, in remote-controlled mode, a certain number of predetermined points stored in the inertial unit of the aircraft.

Sometimes, the destruction of the aircraft is required and it is necessary to recompose the trajectories telemeasured and/or determined following orders from the control surfaces order maker, in order to determine the origin of the failures and to make the necessary corrections.

7.1. Simulation model of the atmosphere

If $Z \leq 11{,}000$:

$$T = T_0 - 6.5 * Z * 10^{-3}$$

$$P = 101{,}325 * \left(\frac{T}{T_0}\right)^{5.2559}$$

If $Z > 11{,}000$:

$$T = T_0 - 71.5$$

$$P = P_{11{,}000} * e^{(((-34.16) * ((Z * 10-3) - 11))/T)}$$

where:

Z is the altitude expressed in meters;

P is the pressure expressed in pascals;

T is the temperature expressed in Kelvin.

7.2. Propulsive coefficients

The engine model makes it possible to calculate the thrust F and the hourly consumption C as a function of the number of revolutions N and the Mach number M.

7.3. Mass coefficients

During a flight, the emptying of the kerosene tanks is always done in the same order (e.g. front tank then rear tank and finally central tank).

It is therefore possible, knowing the instantaneous mass, to deduce the position of the center of gravity and the moments of inertia with respect to the main axes of inertia.

The instantaneous mass is given by the following equation:

$$M = M_0 - \int_0^t \left(\frac{CH}{3600}\right) * dt$$

where:

M_0 is the mass at the origin of the aircraft;

M is the instantaneous mass;

CH is the hourly consumption.

7.4. Aerodynamic coefficients

Aerodynamic coefficients are used to model aerodynamic forces and the moments they create with respect to the center of gravity.

The forces are given by the coefficients:

– $C_x = C_{x0} + [k + \lambda * (X_R - X_G)] * C_N^2$ (in aerodynamic speed axes);

– $C_y = C_{y0} + C_{N\varsigma}(M, \eta_{ex}) * \varsigma_{ex} + \left[C_{y\beta}(M) + C_{y\beta\eta}(M) * \eta_{ex}\right] * \beta + C_{N\xi}(M, \eta_{ex}) * \xi_{ex} + C_{Yr}(M, X_G) * l * r/V$ (in aircraft axes);

– $C_N = C_N(M, \eta_{ex}) + C_{N\alpha}(M) * \alpha + C_{Nq}(M, X_G) * l * \frac{q}{V}$ (in aircraft axes).

We can write, in aircraft axes:

$$\vec{R} = \begin{pmatrix} R_T \\ R_Y \\ R_N \end{pmatrix} = \begin{pmatrix} 0.7 * P * M^2 * S * C_X - R_N * \sin\alpha \\ 0.7 * P * M^2 * S * C_y \\ 0.7 * P * M^2 * S * C_N = n * m * g \end{pmatrix}$$

The drag R_T is > 0 backwards, the lateral force R_Y is > 0 to the right and the lift force R_N is > 0 upwards.

The moments with respect to a reference point R used in a wind tunnel are given by:

$$C_l = C_{l0} + C_{l\xi}(M, \eta_{ex}) * \xi_{ex} + C_{lp}(M) * l * \frac{P}{V} + [C_{l\beta}(M) + Cl_{y\beta\alpha}(M) * \alpha] * \beta$$

$$C_m = C_m(M, \eta_{ex}) + C_{mq}(M, X_G) * l * q/V + [C_{m\alpha}(M) + C_{m\alpha\beta}(M) * |\beta|] * \alpha$$

$$C_n = C_{no} + C_{N\varsigma}(M, \eta_{ex}) * \varsigma_{ex} + C_{nr}(M, X_G) * l * \frac{r}{V} + C_{n\xi}(M, \eta_{ex}) * \xi_{ex} + \left[C_{n\beta}(M) + C_{n\beta\eta}(M) * \eta_{ex}\right] * \beta$$

This reference point having been chosen on the longitudinal O_x axis of the aircraft at a distance X_r from the front point, the moments relative to the center of gravity are then written as:

$$\overrightarrow{M_G} = \begin{pmatrix} L_X \\ L_Y \\ L_Z \end{pmatrix} = \begin{pmatrix} 0.7 * P * M^2 * S * l * C_l + R_N * Y_G + R_Y * Y_G \\ 0.7 * P * M^2 * S * l * C_m + R_T * Z_G - R_N * (X_R - X_G) + F * (Z_F - Z_G) \\ 0.7 * P * M^2 * S * l * C_l - R_Y * (X_R - X_G) - R_Y * Y_G \end{pmatrix}$$

with:

L_X: the roll moment is > 0 when the aircraft moves to the right;

L_Y: the pitch moment is > 0 when the aircraft pitches upward;

L_Z: the yaw moment is > 0 when the nose of the aircraft goes to the right.

7.5. Equations of motion

The equations of motion are written as follows:

$$\frac{\partial \overrightarrow{V_a}}{\partial t} = \vec{g} + \frac{\vec{F} + \vec{R}}{m} + \overrightarrow{V_a} \wedge \overrightarrow{\Omega_{E/A}} \quad [7.1]$$

$$\frac{\partial (I_G * \overrightarrow{\Omega_{E/A}})}{\partial t} = \overrightarrow{M_G} \quad [7.2]$$

where $\overrightarrow{V_a}$ is the absolute speed and $\overrightarrow{\Omega_{E/A}}$ the instantaneous speed vector of rotation of the aircraft trihedron with respect to an absolute reference A.

Since the drone is not symmetrical with respect to the YGX plane, the inertia tensor is written as:

$$I_G = \begin{pmatrix} I_x & 0 & -I_{xz} \\ 0 & I_y & 0 \\ -I_{xz} & 0 & I_z \end{pmatrix}$$

We can note:

– the components of $\vec{V_a}$ in the aircraft trihedron E:

$$\begin{pmatrix} u \\ v \\ w \end{pmatrix}_E$$

– the components of $\overrightarrow{\Omega_{E/A}}$ in the aircraft trihedron E:

$$\begin{pmatrix} p \\ q \\ r \end{pmatrix}_E$$

– the Euler angles defining the position of the aircraft trihedron E with respect to the local geographic triad L in which the vector $\vec{g}$ has, for components (0, 0, g):

$$(\Psi, \theta, \Phi)$$

From equation [7.1]:

$$\frac{\partial \overrightarrow{V_a}}{\partial t} = \vec{g} + \frac{\vec{F} + \vec{R}}{m}$$

$$u' = g * \sin\theta - w * q + v * r + \frac{F + R_T}{m}$$

$$v' = g * \sin\Phi * \cos\theta - u * r + w * p + \frac{R_Y}{m}$$

$$w' = g * \cos\Phi * \cos\theta - v * p + u * a - \frac{R_z}{m}$$

and:

$$\frac{\partial(I_G * \overrightarrow{\Omega_{E/A}})}{\partial t} = \overrightarrow{M_G}$$

$$p' = \frac{1}{(I_x * I_z - I_{xz}^2)} * (I_{xz} * (I_x - I_y + I_z) * p * q - (I_z^2 + I_{xz}^2 - I_y * I_z) * q * r + I_z * L + I_{xz} * N)$$

$$q' = \frac{1}{I_y} * ((I_z - I_x) * p * r + I_{xz} * (r^2 - p^2) + M)$$

$$r' = \frac{1}{(I_x * I_z - I_{xz}^2)} * ((I_x^2 + I_{xz}^2 - I_x * I_y) * p * q - I_{xz} * (I_x - I_y + I_z) * q * r + I_x * N + I_{xz} * L)$$

The integration of these differential equations with a numerical integration method makes calculations possible at each calculation step u, v, w, p, q and r.

The instantaneous speed vector of rotation of the aircraft trihedron E with respect to the local geographical trihedron (LGT) can be written in aircraft axes:

$$\vec{\Omega}_{E/LGT} = \begin{pmatrix} \Phi' - \psi' * \sin\theta \\ \theta' * \cos\Phi + \psi' * \cos\theta * \sin\Phi \\ -(\theta' * \sin\Phi) + \psi' * \cos\theta * \cos\Phi \end{pmatrix}_E$$

Likewise, the instantaneous speed vector of rotation of the local geographical frame L with respect to the absolute frame of reference A can be written as:

$$\vec{\Omega}_{LGT/A} = \begin{pmatrix} p_{LGT} \\ q_{LGT} \\ r_{LGT} \end{pmatrix}_{LGT}$$

with:

$$p_{LGT} = -\lambda_M' * \sin\psi * \cos\theta + (G_M' + \Omega_{Earth}) * [\cos(\lambda_{CT} + \lambda_M) * \cos\psi * \cos\theta + \sin(\lambda_{CT} + \lambda_M) * \sin\theta]$$

$$q_{LGT} = -\lambda'_M * [\cos\psi * \cos\Phi + \sin\psi * \sin\theta * \sin\Phi] + (G'_M + \Omega_{Earth}) * [\cos(\lambda_{CT} + \lambda_M) * (\cos\psi * \cos\theta * \sin\Phi - \sin\psi * \cos\Phi) - \sin(\lambda_{CT} + \lambda_M) * \cos\theta * \sin\Phi]$$

$$r_{LGT} = -\lambda'_M * [\sin\psi * \sin\theta * \cos\Phi - \cos\psi * \sin\Phi] + (G'_M + \Omega_{Earth}) * [\cos(\lambda_{CT} + \lambda_M) * (\cos\psi * \sin\theta * \cos\Phi + \sin\psi * \sin\Phi) - \sin(\lambda_{CT} + \lambda_M) * \cos\theta * \cos\Phi]$$

and where:

Ω_{Earth} is the Earth's speed of rotation;

λ_M is the difference between the latitude of the aircraft and the latitude of the origin of the test runway;

λ_{CT} is the latitude of the test runway origin;

G_M is the deviation between the longitude of the aircraft and the longitude of the origin of the test runway.

By writing:

$$\vec{\Omega}_{E/A} = \vec{\Omega}_{E/LGT} + \vec{\Omega}_{LGT/A}$$

we end up with the following equations:

$$\psi' = [(q - q_L) * \sin\Phi + (r - r_L) * \cos\Phi] / \cos\theta$$

$$\theta' = (q - q_L) * \cos\Phi - (r - r_L) * \sin\Phi$$

$$\Phi' = (p - p_L) + \psi' * \sin\theta$$

The integration provides us with the attitudes ψ, θ and φ of the aircraft with respect to the local geographic coordinate system (the LGT), hence the trihedron change matrix (the reference trihedron is the trihedron defined at the point of projection of the center of gravity of the drone on the test runway (O, X_0, Y_0, Z_0); see section 4.5.1):

$$M_{E/LGT} = \begin{pmatrix} M_{11} & M_{12} & M_{13} \\ M_{21} & M_{22} & M_{23} \\ M_{31} & M_{32} & M_{33} \end{pmatrix}_{E/LGT}$$

with:

$$M_{11} = \cos\Psi * \cos\theta$$

$$M_{12} = \sin\Psi * \cos\theta$$

$$M_{13} = -\sin\theta$$

$$M_{21} = ((-\sin\Psi * \cos\Phi) + (\cos\Psi * \sin\theta * \sin\Phi))$$

$$M_{22} = ((\cos\Psi * \cos\Phi) + (\sin\Psi * \sin\theta * \sin\Phi))$$

$$M_{23} = \cos\theta * \sin\Phi$$

$$M_{31} = ((\sin\Psi * \sin\Phi) + (\cos\Psi * \sin\theta * \cos\Phi))$$

$$M_{32} = ((-\cos\Psi * \sin\Phi) + (\sin\Psi * \sin\theta * \cos\Phi))$$

$$M_{33} = \cos\theta * \cos\Phi$$

The relative speed of the aircraft with respect to the Earth is given by:

$$\vec{V}_{Earth}(M) = \vec{V}_a(M) + \vec{\Omega}_{Earth} \wedge \overrightarrow{OM}$$

with the Earth reference:

$$\vec{\Omega}_{Earth} = \begin{pmatrix} 0 \\ 0 \\ \Omega_{Earth} \end{pmatrix}_{Earth}$$

and:

$$\overrightarrow{OM} = \begin{pmatrix} (R + H_M) * \cos(\lambda_{CT} + \lambda_M) * \cos(G_M) \\ (R + H_M) * \cos(\lambda_{CT} + \lambda_M) * \sin(G_M) \\ (R + H_M) * \sin(\lambda_{CT} + \lambda_M) \end{pmatrix}_{Earth}$$

where R is the radius of the Earth and H_M is the aircraft altitude.

From these equations, it follows that:

$$\overrightarrow{\Omega}_{Earth} \wedge \overrightarrow{OM} = \begin{pmatrix} (-\Omega_{Earth}) * (R + H_M) * \cos(\lambda_{CT} + \lambda_M) * \sin(G_M) \\ \Omega_{Earth} * (R + H_M) * \cos(\lambda_{CT} + \lambda_M) * \cos(G_M) \\ 0 \end{pmatrix}_{Earth}$$

The change of reference matrix making it possible to go from the Earth reference frame to the local geographic reference frame is as follows:

$$M_{LGT/Earth} = \begin{pmatrix} M_{11} & M_{12} & M_{13} \\ M_{21} & M_{22} & M_{23} \\ M_{31} & M_{32} & M_{33} \end{pmatrix}_{LGT/Earth}$$

with:

$$M_{11} = (-\sin(\lambda_{CT} + \lambda_M)) * \cos(G_M)$$

$$M_{12} = (-\sin(\lambda_{CT} + \lambda_M)) * \sin(G_M)$$

$$M_{13} = \cos(\lambda_{CT} + \lambda_M) * \sin(G_M)$$

$$M_{21} = -\sin(G_M)$$

$$M_{22} = \cos(G_M)$$

$$M_{23} = 0$$

$$M_{31} = (-\cos(\lambda_{CT} + \lambda_M)) * \cos(G_M)$$

$$M_{32} = (-\cos(\lambda_{CT} + \lambda_M)) * \sin(G_M)$$

$$M_{33} = (-\sin(\lambda_{CT} + \lambda_M))$$

Hence, calculating the components of the relative speed vector with respect to the Earth in the local geographic reference frame (descending vertical):

$$\begin{pmatrix} V_N \\ V_E \\ V_Z \end{pmatrix}_{LGT} = M^{-1}_{\frac{E}{LGT}} * \begin{pmatrix} u \\ v \\ w \end{pmatrix}_E - M_{\frac{LGT}{Earth}}$$

$$* \begin{pmatrix} (-\Omega_{Earth}) * (R + H_M) * \cos(\lambda_{CT} + \lambda_M) * \sin(G_M) \\ \Omega_{Earth} * (R + H_M) * \cos(\lambda_{CT} + \lambda_M) * \cos(G_M) \\ 0 \end{pmatrix}_{Earth}$$

we obtain:

$$\lambda'_M = \frac{V_N}{R + H_M}$$

$$G'_M = \frac{V_E}{(R + H_M) * \cos(\lambda_{CT} + \lambda_M)}$$

and:

$$\lambda_M = \lambda_0 + \int_0^t \lambda'_M * dt \text{ (given in radians)}$$

$$G_M = G_0 + \int_0^t G'_M * dt \text{ (given in radians)}$$

$$H_M = H_0 + \int_0^t V_Z * dt$$

Hence:

– λ0 is the deviation in latitude of the aircraft's center of gravity from the origin of the test runway;

– G0 is the deviation in longitude of the aircraft's center of gravity from the origin of the test runway.

7.6. Introduction of wind

Wind can be introduced into the simulation. It is defined by its modulus V_v and by its heading C_v with respect to the local geographic north. In the trihedron (the LGT), this wind is expressed as follows:

$$\vec{V}_v = \begin{pmatrix} V_v * \cos C_v \\ V_v * \sin C_v \\ 0 \end{pmatrix}_{LGT}$$

The aerodynamic speed in the local geographical reference is then given by:

$$\vec{V}_{ae} = \vec{V}_R - \vec{V}_v$$

and has the following for components in the aircraft trihedron:

$$\begin{pmatrix} u_{ae} \\ v_{ae} \\ w_{ae} \end{pmatrix}_E = M_{\frac{E}{LGT}} * \begin{pmatrix} V_N - V_v * \cos C_v \\ V_E - V_v * \sin C_v \\ V_Z \end{pmatrix}_E$$

$$= M_{E/LGT} * \begin{pmatrix} V_N - V_{vnorth} \\ V_E - V_{veast} \\ V_Z \end{pmatrix}_E$$

Hence, the incidence is given by:

$$\alpha = \text{Arctg}(w_{ae}/u_{ae})$$

The aerodynamic speed is given by:

$$V_{ae} = \sqrt{(u_{ae}^2 + v_{ae}^2 + w_{ae}^2)}$$

The skid is given by:

$$\beta = \text{Arcsin}(v_{ae}/V_{ae})$$

and the Mach number is given by (see section 6.1.4):

$$M = V_{ae}/V_{sound} = V_{ae}/[20{,}0807 * \sqrt{T}]$$

We thus obtain the aerodynamic forces and torques.

8

Processing of Trajectography Information

Here we show, for an aircraft with aerodynamic wings and with the aerodynamics being defined in aircraft axes, that it is possible to verify, using the trajectory and the information obtained from telemetry, the performance and some major aerodynamic parameters specific to the aircraft studied.

8.1. Telemeasured parameters

The elimination of outliers must be undertaken with the raw coordinates (x_c, y_c, z_c) or z_{geoid}.

This operation should be undertaken seperately, because it gives us an idea of the quality of the trajectography.

8.2. Smoothing, first derivation and second derivation

For processing from the trajectography information, it is preferable to use polynomial smoothing with first derivatives and second derivatives. This processing will be carried out with the coordinates (x_c, y_c, z_c), after possible realignment $(\Delta x_c, \Delta y_c, \Delta z_c)$ with respect to the origin of the trihedron.

If we have the value z_{geoid}, we can take:

$$z_c = z_{geoid} - \frac{(x_c - \Delta x_c)^2 + (y_c - \Delta y_c)^2}{2 * R}$$

We then obtain:

– x_c, y_c, z_c: smoothed coordinates;

– $\dot{x}_c, \dot{y}_c, \dot{z}_c$: smoothed speeds;

– $\ddot{x}_c, \ddot{y}_c, \ddot{z}_c$: smoothed acceleration.

8.3. Performance calculation

8.3.1. *Change of coordinates to move from the trihedron (O, X_c, Y_c, Z_c) to the trihedron (O, X_0, Y_0, Z_0)*

(O, X_0, Y_0, Z_0) is the reference trihedron used for the calculation. The angles Ψ_{0c} and Φ_{0c} depend on the trihedrons of restitution of the trajectories used for the different test centers.

In general:

$$\Psi_{0c} = k * (\pi/2) \text{ and } \Phi_{0c} = \pi$$

We move from the trihedron (O, X_0, Y_0, Z_0) to the trihedron (O, X_c, Y_c, Z_c) with the two following rotations:

$$(O, X_c, Y_c, Z_c) \rightarrow (O, X_0, Y_1, Z_c) \text{ by rotation } \Psi_{0c} \text{ around } (O, Z_c)$$

$$(O, X_0, Y_1, Z_c) \rightarrow (O, X_0, Y_0, Z_0) \text{ by rotation } \Psi_{0c} \text{ around } (O, X_0)$$

$$\begin{pmatrix} I_c \\ J_c \\ Z_c \end{pmatrix} = \begin{pmatrix} \cos\Psi_{0c} & -\sin\Psi_{0c} & 0 \\ \sin\Psi_{0c} & \cos\Psi_{0c} & 0 \\ 0 & 0 & 1 \end{pmatrix} * \begin{pmatrix} 1 & 0 & 0 \\ 0 & \cos\Phi_{0c} & -\sin\Phi_{0c} \\ 0 & \sin\Phi_{0c} & \cos\Phi_{0c} \end{pmatrix} * \begin{pmatrix} I_0 \\ J_0 \\ Z_0 \end{pmatrix}$$

$$\begin{pmatrix} I_c \\ J_c \\ Z_c \end{pmatrix} = \begin{pmatrix} \cos\Psi_{0c} & -\sin\Psi_{0c} * \cos\Phi_{0c} & \sin\Psi_{0c} * \sin\Phi_{0c} \\ \sin\Psi_{0c} & \cos\Psi_{0c} * \cos\Phi_{0c} & -\cos\Psi_{0c} * \sin\Phi_{0c} \\ 0 & \sin\Phi_{0c} & \cos\Phi_{0c} \end{pmatrix} * \begin{pmatrix} I_0 \\ J_0 \\ Z_0 \end{pmatrix}$$

$$\begin{pmatrix} I_0 \\ J_0 \\ Z_0 \end{pmatrix} = \begin{pmatrix} \cos\Psi_{0c} & \sin\Psi_{0c} & 0 \\ -\sin\Psi_{0c} * \cos\Phi_{0c} & \cos\Psi_{0c} * \cos\Phi_{0c} & \sin\Phi_{0c} \\ \sin\Psi_{0c} * \sin\Phi_{0c} & -\cos\Psi_{0c} * \sin\Phi_{0c} & \cos\Phi_{0c} \end{pmatrix} * \begin{pmatrix} I_c \\ J_c \\ Z_c \end{pmatrix}$$

We thus obtain the coordinates (x_0, y_0, z_0) as well as their first and second derivatives.

The geoid position z_{0geoid} is given by the following equation:

$$z_{0geoid} = z_0 - \frac{x_0^2 + y_0^2}{2 * R}$$

8.3.2. *Aircraft speed relative to the ground*

The speed of the aircraft relative to the ground is given by the following formula:

$$V^2 = \frac{\delta x_0}{\delta t} + \frac{\delta y_0}{\delta t} + \frac{\delta z_0}{\delta t}$$

8.3.3. *Aerodynamic speed and Mach number*

V_v and C_v depend on the local altitude H ($H = -z_{0geoid}$).

The wind speed breaks down as follows:

$$\vec{V}_v = u_v * \vec{I}_0 + v_v * \vec{J}_0 + w_v * \vec{K}_0$$

The aerodynamic speed is given by:

$$\vec{V}_a = \vec{V} - \vec{V}_v$$

The wind is given in terms of speed and direction (direction facing the wind).

We have:

$$\chi_v = C_v + \pi$$

and:

$$u_v = V_v * \cos\chi_v = -V_v * \cos(C_v + \pi) = -V_v * \cos(C_v)$$

$$v_v = V_v * \sin\chi_v = -V_v * \sin(C_v + \pi) = -V_v * \sin(C_v)$$

$$w_v = 0$$

The components of the aerodynamic speed in the trihedron (O, X_0, Y_0, Z_0) are:

$$u_0 = \frac{\delta x_0}{\delta t} + V_v * \cos C_v$$

$$u_0 = \frac{\delta y_0}{\delta t} + V_v * \sin C_v$$

$$u_0 = \frac{\delta z_0}{\delta t}$$

The aerodynamic speed is therefore:

$$|V_a| = \sqrt{(u_0^2 + v_0^2 + w_0^2)}$$

The Mach number is given by:

$$M = \frac{V_a}{20.0807 \ * \ \sqrt{T}}$$

At the same time, we can calculate:

$$Q_S = 0.7 * P * M^2 * S$$

If we have neither the temperature nor the pressure, we can use the standard atmosphere:

$$T = T_0 - 0.0065 * (H - H_0)$$

$$P = P_0 * \frac{|T|^{5,2558774}}{|T_0|}$$

P_0 and T_0 are the pressure and temperature at altitude H_0, respectively.

8.4. Aerodynamic route and slope

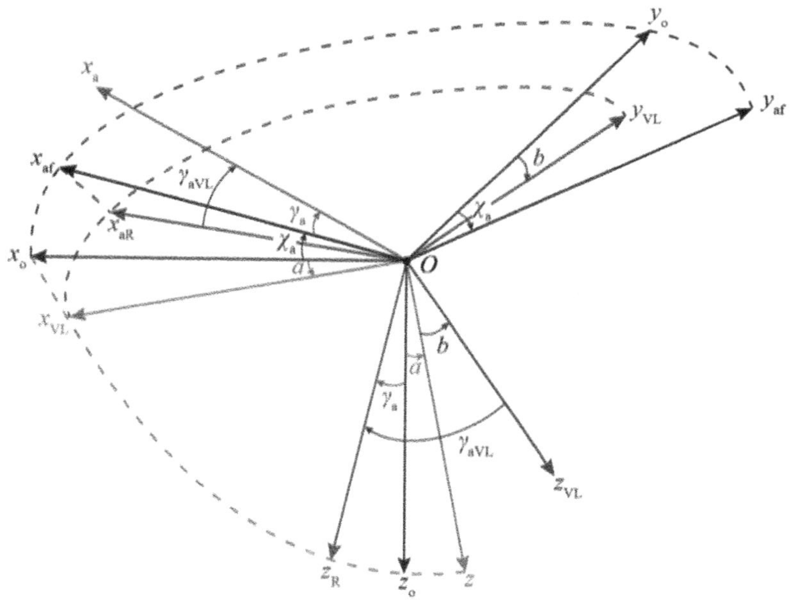

Figure 8.1. *Trihedron related to the route and the aerodynamic slope. The points (X_O, X_{aF}, Y_O, Y_{aF}) are in the same plane; X_{aR} is the projection of X_{aF} on the plane (O, X_{VL}, Y_{VL}); the points (X_{VL}, X_{aR}, Y_{VL}) are in the same plane. For a color version of this figure, see www.iste.co.uk/louis/flight.zip*

Let the trihedron of reference computation (O, X_0, Y_0, Z_0) be defined as follows:

– the axis $\overrightarrow{OX_0}$ is directed north;

– the axis $\overrightarrow{OZ_0}$ is directed downwards (center of the Earth);

– the axis $\overrightarrow{OY_0}$ completes this direct trihedron.

The trihedron related to the local vertical $(O, X_{vl}, Y_{vl}, Z_{vl})$ is defined as follows:

– the axis $\overrightarrow{OX_{vl}}$ is parallel to the calculation plane $(O,\ X_0,\ Y_0)$;

– the axis $\overrightarrow{OZ_{vl}}$ is directed downwards along the local vertical (center of the Earth);

– the axis $\overrightarrow{OY_{vl}}$ completes this direct trihedron.

The aerodynamic trihedron of the aircraft (O, X_a, Y_{af}, Z_r) is defined as follows:

– the axis $\overrightarrow{OX_a}$ is parallel to the calculation plane $(OX_0,\ OY_0)$;

– the axis $\overrightarrow{OY_{af}}$ is directed downwards along the local vertical (center of the Earth);

– the axis $\overrightarrow{OZ_r}$ completes this direct trihedron.

In the trihedron $(O,\ X_0,\ Y_0,\ Z_0)$, the aerodynamic route is given by the relation:

$$\chi_a = \text{Arctg} * \frac{|V_0|}{|U_0|}$$

If we take the route in relation to any direction (e.g. the axis of the aircraft), we can write:

$$\chi_{ac} = \chi_a - \psi 0$$

On the other hand, the aerodynamic slope is given by:

$$\gamma a = \text{Arctg} * \left(\frac{-w_0}{\sqrt{(u_0^2 + v_0^2)}}\right) = \text{Arctg} * \left(\frac{-w_0}{V_a}\right)$$

with:

$$V_a = \sqrt{(u_0^2 + v_0^2 + w_0^2)}$$

(u_0, v_0, w_0) are the components of the aerodynamic speed in the ground coordinate system.

To pass from the trihedron (O, X_0, Y_0, Z_0) to the trihedron $(O, X_{vl}, Y_{vl}, Z_{vl})$, we carry out the following two successive rotations:

$(O, X_0, Y_0, Z_0) \rightarrow (O, X_{vl}, Y_0, Z)$ by rotation a around (O, Y_o)

$(O, X_{vl}, Y_0, Z) \rightarrow (O, X_{vl}, Y_{vl}, Z_{vl})$ by rotation b around (O, X_{vl})

and we can write:

$$a = \frac{x_0}{R} \text{ and } a = \frac{x_0}{R}$$

However, a and b are very small angles:

$$\begin{pmatrix} I_{vl} \\ J_{vl} \\ K_{vl} \end{pmatrix} = \begin{pmatrix} 1 & 0 & 0 \\ 0 & \cos b & \sin b \\ 0 & -\sin b & \cos b \end{pmatrix} * \begin{pmatrix} \cos a & 0 & \sin a \\ 0 & 1 & 0 \\ -\sin a & 0 & \cos a \end{pmatrix} * \begin{pmatrix} I_0 \\ J_0 \\ Z_0 \end{pmatrix}$$

$$\begin{pmatrix} I_{vl} \\ J_{vl} \\ K_{vl} \end{pmatrix} = \begin{pmatrix} \cos a & 0 & \sin a \\ -\sin a * \sin b & \cos b & \sin b * \cos a \\ -\cos b * \sin a & -\sin b & \cos b * \cos a \end{pmatrix} * \begin{pmatrix} I_0 \\ J_0 \\ Z_0 \end{pmatrix}$$

With the small angle approximation, we can simplify the matrix as follows:

$$\begin{pmatrix} I_0 \\ J_0 \\ Z_0 \end{pmatrix} = \begin{pmatrix} 1 & 0 & a \\ 0 & 1 & b \\ -a & -b & 1 \end{pmatrix} * \begin{pmatrix} I_c \\ J_c \\ Z_c \end{pmatrix}$$

The component of the speed vector of the aircraft on the axis of the local vertical is then:

$$\delta z_{vl} = \frac{-\left(\frac{\delta x_0}{\delta t} * a + \frac{\delta y_0}{\delta t} * a\right)}{R} + \frac{\delta z_0}{\delta t}$$

$$\delta z_{vl} = \frac{-\left(\frac{\delta x_0}{\delta t} * x_0 - \frac{\delta y_0}{\delta t} * y_0\right)}{R} + \frac{\delta z_0}{\delta t} \quad \text{verification of the sign of yO}$$

$$w_{vl} = \frac{-(u_0 * x_0) + v_0 * y_0}{R} + w_0 \quad \text{verification of the sign}$$

If we refer to the local vertical, the local slope is given by:

$$\gamma a = \text{Arctg}\left(-\frac{w_0 - \frac{(u_0 * x_0) + v_0 * y_0}{R}}{\sqrt{\left(u_0 + \frac{w_0 * x_0}{R}\right)^2 + \left(v_0 + \frac{w_0 * y_0}{R}\right)^2}} \right)$$

$$\gamma a = \text{Arcsin}\left(\frac{w_0 - \frac{(u_0 * x_0) + v_0 * y_0}{R}}{V_a} \right)$$

8.5. Determination of the angle ψ

Knowing the components of the speed in the ground coordinate system, or local Galilean coordinate system, makes it possible to deduce the angle Ψ at any time.

In fact, the angle Ψ represents the angle between the east speed and the north speed. Therefore, this angle is given by the following relation:

$$\Psi = \text{Arctg}\left(\frac{V_{east}}{V_{north}}\right)$$

In fact, we can assume that the heading is equal to the course χ of the aircraft. It is the route χ which is equal to $\text{Arctg}(V_{east}/V_{north})$. This assumption is all the more justified since there is no crosswind and the aircraft flies at low sideslip angles.

8.6. Load factors in the aerodynamic trihedron (O, X_a, Y_{af}, Z_r)

To pass from the trihedron (O, X_a, Y_{af}, Z_r) to the trihedron (O, X_0, Y_0, Z_0), defined in section 8.4, we carry out two successive rotations:

$(O, X_0, Y_0, Z_0) \rightarrow (O, X_{af}, Y_{af}, Z_0)$ by rotation χ_a around $\overrightarrow{OZ_0}$

$(O, X_{af}, Y_{af}, Z_0) \rightarrow (O, X_a, Y_{af}, Z_r)$ by rotation γ_a around $\overrightarrow{OY_{af}}$

We have:

$$\overrightarrow{OX_0} = \cos\chi_a * \overrightarrow{OX_{af}} - \sin\chi_a * \overrightarrow{OY_{af}}$$

$$\overrightarrow{OY_0} = \sin\gamma_a * \overrightarrow{OX_{af}} + \cos\gamma_a * \overrightarrow{OY_{af}}$$

$$\begin{pmatrix} I_0 \\ J_0 \\ K_0 \end{pmatrix} = \begin{pmatrix} \cos\chi_a & -\sin\chi_a & 0 \\ \sin\chi_a & \cos\chi_a & 0 \\ 0 & 0 & 1 \end{pmatrix} * \begin{pmatrix} \cos\gamma_a & 0 & \sin\gamma_a \\ 0 & 1 & 0 \\ -\sin\gamma_a & 0 & \cos\gamma_a \end{pmatrix} * \begin{pmatrix} I_a \\ J_{af} \\ K_r \end{pmatrix}$$

$$\begin{pmatrix} I_0 \\ J_0 \\ K_0 \end{pmatrix} = \begin{pmatrix} \cos\chi_a * \cos\gamma_a & -\sin\chi_a & \cos\chi_a * \sin\gamma_a \\ \sin\chi_a * \cos\gamma_a & \cos\chi_a & \sin\chi_a * \sin\gamma_a \\ -\sin\gamma_a & 0 & \cos\gamma_a \end{pmatrix} * \begin{pmatrix} I_a \\ J_{af} \\ K_r \end{pmatrix}$$

$$\begin{pmatrix} I_a \\ J_{af} \\ K_r \end{pmatrix} = \begin{pmatrix} \cos\chi_a * \cos\gamma_a & \sin\chi_a * \cos\gamma_a & -\sin\gamma_a \\ -\sin\chi_a & \cos\chi_a & 0 \\ \cos\chi_a * \sin\gamma_a & \sin\chi_a * \sin\gamma_a & \cos\gamma_a \end{pmatrix} * \begin{pmatrix} I_0 \\ J_0 \\ K_0 \end{pmatrix}$$

The load factors in the reference (O, X_a, Y_{af}, Z_r) are given by the following equations:

$$n_{xa} = \frac{\frac{\delta^2 x_a}{\delta^2 t}}{g} + \sin\gamma_{avl}$$

$$n_{xa} = \frac{\frac{\delta^2 y_a}{\delta^2 t}}{g}$$

$$n_{xa} = \frac{\frac{\delta^2 z_a}{\delta^2 t}}{g} - \cos\gamma_{avl}$$

8.7. Processing of data from the inertial unit (in the aircraft frame)

Note that C is the point where the inertial unit is located (providing the measurements taken on board the aircraft) and G is the center of gravity of the aircraft.

The position of the take-off runway of the aircraft in relation to the local geographic coordinate system is known; we can take it to be equal to the center of gravity of the aircraft at the time of the test.

The coordinates of the inertial unit in relation to the center of gravity of the aircraft are known.

The telemetry is carried out at the level of the aircraft inertial unit and sent back to the ground station.

If we have, using telemetry, the attitude, the roll, the coefficients of the speed of rotation of the aircraft (p, q and r) in the local geographic reference, the load factors, the steering ordered or performed and the data from the radar, it is possible to recalculate the flight trajectory and compare it with that recorded by the radar.

The difference between the probationary and recalculated trajectories makes it possible to calculate the differences between the aerodynamic coefficients.

The coefficients p', q' and r' are generally obtained by filtering, derivation and smoothing.

At the instant t = 0 s, we know the position of the take-off runway, which is equal to the position of the center of gravity, itself known with respect to the local geographical reference.

At the first telemetry step, knowing the parameters makes it possible to undertake all the calculations.

The accelerations are given by the difference between the speed at the origin (equal to 0 m/s) and that at step = 1 divided by the step time-interval.

8.7.1. *Load factor conversion*

$$\vec{V}_{LGT}(G) = \left(\frac{d\overrightarrow{OG}}{dt}\right)_{LGT}$$

$$\vec{V}_{LGT}(C) = \left(\frac{d\overrightarrow{OG}}{dt}\right)_{LGT} + \left(\frac{d\overrightarrow{GC}}{dt}\right)_{LGT} = \vec{V}_{LGT}(G) + \vec{\Omega}_{E/LGT}{}^\wedge\overrightarrow{GC}$$

$$\left(\frac{d\overrightarrow{GC}}{dt}\right)_{LGT} = \vec{V}_{LGT}(G) - \vec{V}_{LGT}(C) = \vec{\Omega}_{E/LGT}{}^\wedge\overrightarrow{GC}$$

Thus:

$$\vec{V}(G) = \vec{V}(C) - \vec{\Omega} \wedge \overrightarrow{GC}$$

In the LGT, we have:

$$\left(\frac{d\vec{V}(G)}{dt}\right)_{LGT} = \left(\frac{d\vec{V}(C)}{dt}\right)_{LGT} - \vec{\Omega}_{E/LGT}{}^\wedge\left(\frac{d\overrightarrow{GC}}{dt}\right) - \left(\frac{d\vec{\Omega}_{E/LGT}}{dt}\right)^\wedge$$

$$\vec{\Gamma}_{LGT}(G) = \vec{\Gamma}_{LGT}(C) - \left(\frac{d\vec{\Omega}_{E/LGT}}{dt}\right) \wedge \overrightarrow{GC} - \vec{\Omega}_{E/LGT}{}^\wedge(\vec{\Omega}_{E/LGT}{}^\wedge\overrightarrow{GC})$$

According to the equations in section 5.7, the load factors at the center of gravity can be written in aircraft axes using the load factors measured at the inertial unit, as follows:

$$\begin{pmatrix} n_{xG} \\ n_{yG} \\ n_{zG} \end{pmatrix} = \begin{pmatrix} n_{xC} \\ n_{yC} \\ n_{zC} \end{pmatrix} - \left(\begin{pmatrix} p' \\ q' \\ r' \end{pmatrix} \wedge \begin{pmatrix} X_G - X_C \\ Y_C - Y_G \\ Z_C - Z_G \end{pmatrix}\right)$$
$$- \left(\begin{pmatrix} p \\ q \\ r \end{pmatrix} \wedge \left(\begin{pmatrix} p \\ q \\ r \end{pmatrix} \wedge \begin{pmatrix} X_G - X_C \\ Y_C - Y_G \\ Z_C - Z_G \end{pmatrix}\right)\right)$$

$X_G - X_C$ because X_G and X_C are >0 with respect to the nose cone of the aircraft.

By noting:

$$\begin{pmatrix} X_G - X_C \\ Y_C - Y_G \\ Z_C - Z_G \end{pmatrix} = \begin{pmatrix} \Delta X \\ \Delta Y \\ \Delta Z \end{pmatrix}$$

n_{xG} becomes:

$$n_{xG} = n_{xC} - q' * (Z_C - Z_G) + r' * (Y_C - Y_G) - q * (p * (Y_C - Y_G) - q * (X_G - X_C)) + r * (r * (X_G - X_C) - p * (Z_C - Z_G))$$

We obtain:

$$n_{xG} = n_{xC} - q' * \Delta Z + r' * \Delta Y - q * p * \Delta Y + q^2 * \Delta X + r^2 * \Delta X - p * r * \Delta Z$$

$$n_{yG} = n_{yC} - r' * \Delta X + p' * \Delta Z + (p^2 + r^2) * \Delta Y - q * r * \Delta Z + p * q * \Delta X$$

$$n_{zG} = n_{zC} - p' * \Delta Y + q' * \Delta X + (p^2 + q^2) * \Delta Z - p * r * \Delta X - q * r * \Delta Y$$

8.7.2. *Calculation of accelerations at the center of gravity*

The accelerations at the center of gravity in the local geographical trihedron are denoted as $(A_{NG}, A_{EG}, A_{ZG})_{LGT}$. In the aircraft trihedron, knowing the load factors telemeasured in the local geographical reference, and following section 6.2, these accelerations are given by the following relation:

$$\begin{pmatrix} A_{NG} \\ A_{EG} \\ A_{ZG} \end{pmatrix}_E = \begin{pmatrix} n_{xG} * g \\ n_{yG} * g \\ n_{zG} * g \end{pmatrix}_E$$

8.7.3. *Speed conversion*

Let (V_{NG}, V_{EG}, V_{ZG}) be the coordinates of the speed vector (north, east, Z) brought to the center of gravity in the aircraft trihedron. The coordinates of the speed vector $(V_{NG}, V_{EG}, V_{ZG})_{LGT}$ of the center of gravity in the LGT trihedron can be calculated thanks to the conversion matrix $M(\theta, \Phi, \Psi)$.

The components p_{lgt}, q_{lgt}, r_{lgt} are the components of the instantaneous speed vector $\vec{\Omega}_{LGT}$ in the base $(O, i_{lgt}, j_{lgt}, k_{lgt})$ in relation to the aircraft reference.

Telemetry gives the speeds of the inertial unit in the local geographic reference point, that is, $(V_{NC}, V_{EC}, V_{ZC})_{LGT}$.

These speeds are identical to the center of gravity in the local geographic reference (conservation of the speed of a solid in motion).

We must first calculate the speed in the local geographic triad at the level of the center of gravity and then use the transit matrices of the local geographic reference frame in the aircraft triad.

According to section 5.3.5.3, we have the relation:

$$\vec{V}_R(M) = \vec{V}_R(O1) + \vec{\Omega}_e \wedge \overrightarrow{O1M} + \vec{V}_{R1}(O1)$$

where O1 is the center of gravity G and M is the position of the inertial unit C.

We can write:

$$\vec{V}_E(G) = \frac{d\overrightarrow{OG}}{dt}$$

$$\vec{V}_E(G) = \left(\frac{d\overrightarrow{OG}}{dt}\right)_E + \left(\frac{d\overrightarrow{GC}}{dt}\right)_{LGT}$$

We know that the speed of the center of gravity in the aircraft frame is zero, so the previous equation reduces to:

$$\vec{V}_E(C) = \vec{V}_E(G) + \vec{\Omega}_{LGT} \wedge \overrightarrow{GC} + \vec{V}_{LGT}(G) = \vec{\Omega}_{LGT} \wedge \overrightarrow{GC} + \vec{V}_{LGT}(G)$$

We can write:

$$\begin{pmatrix} V_{NG} \\ V_{EG} \\ V_{ZG} \end{pmatrix}_E = M(\theta, \Phi, \Psi) * \begin{pmatrix} V_{NC} \\ V_{EC} \\ V_{ZC} \end{pmatrix}_{LGT} + \begin{pmatrix} p_{LGT} \\ q_{LGT} \\ r_{LGT} \end{pmatrix} * \begin{pmatrix} X_C - X_G \\ Y_C - Y_G \\ Z_C - Z_G \end{pmatrix}_E$$

Let:

$$\begin{pmatrix} V_{NG} \\ V_{EG} \\ V_{ZG} \end{pmatrix}_E = M(\theta, \Phi, \Psi) * \begin{pmatrix} V_{NC} \\ V_{EC} \\ V_{ZC} \end{pmatrix}_{LGT} - \begin{pmatrix} p_{LGT} \\ q_{LGT} \\ r_{LGT} \end{pmatrix} * \begin{pmatrix} X_G - X_C \\ Y_G - Y_C \\ Z_G - Z_C \end{pmatrix}_E$$

The matrix $M(\theta, \Phi, \Psi)$ that allows the change of reference from the LGT to the aircraft reference is given in section 4.5.1.

8.7.4. *Calculation of Ψ*

The angle Ψ represents the angle between the east speed and north speed. Knowing these speeds in the local geographic coordinate system, we can calculate this angle at each calculation step in accordance with section 8.5.

8.7.5. *Recomposition of the trajectory*

If we know, at each calculation step, the load factors, angular and linear accelerations, angular and linear speeds and the angles (ψ, θ, φ), it is possible, using numerical integration, to reconstruct the trajectory of the flight and compare it to the trajectory recorded by the radar.

8.8. Determination of some aerodynamic parameters

8.8.1. *Determination of the incidence α and sideslip β*

If we start from the results of the trajectography, α and β can be calculated by iteration.

Indeed, the load factors deduced from the trajectory are restored in the axes of the “aerodynamic speed” trihedron, while the aerodynamic coefficients are given in the axes of the “aircraft trihedron”.

In the iteration, we can also calculate:

– the Euler angles θ and Ψ;

– the load factors n_{x1}, n_{y1} and n_{z1};

– the coefficients C_y and C_z.

At the start of the iteration, we take the α and β obtained in the previous calculation step; for the first calculation step, α and β are taken to be equal to 0.

8.8.2. *Determination of ψac*

8.8.2.1. *Passage of the trihedron (O, X_a, Y_a, Z_a) to the trihedron (O, X_{af}, Y_{af}, Z_{af})*

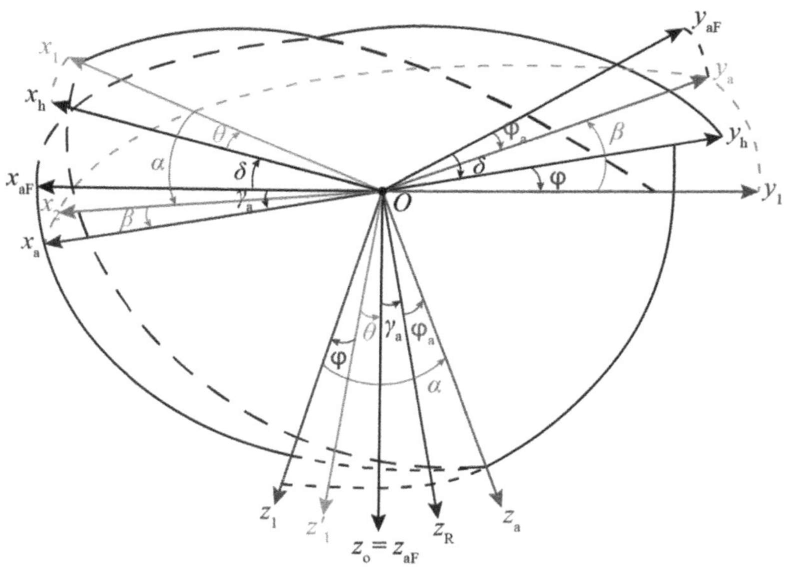

Figure 8.2. *Aerodynamic speed trihedron. The points (X_a, X_2, Y_a, Y_1), (X_{aF}, X_h, Y_{aF}, Y_n), (Y_n, Y_1, Z_1), (X_1, Y_1), (X_1, X_2, Z_a, Z_R), (X_{aF}, X_a, Z_a) and (X_1, X_n, Z_a) are in the same plane. For a color version of this figure, see www.iste.co.uk/louis/flight.zip*

Z_{af} and Z_0 are mixed up.

We go from the trihedron $(0, X_0, Y_0, Z_0)$ to the trihedron $(0, X_{af}, Y_{af}, Z_{af})$ by an angle rotation χ_a around the axis $\overrightarrow{OZ_0}$.

We go from the trihedron $(0, X_{af}, Y_{af}, Z_{af})$ to the trihedron $(0, X_a, Y_a, Z_a)$ by an angle rotation γ_a around the axis $\overrightarrow{OY_{af}}$, followed by an angle rotation Φ_a around the axis $\overrightarrow{OX_a}$.

$$(O, X_{af}, Y_{af}, Z_{af}) \rightarrow (O, X_a, Y_{af}, Z_r) \text{ by rotation } \gamma_a \text{ around } \overrightarrow{OY_{af}}$$

$$(O, X_a, Y_{af}, Z_r) \rightarrow (O, X_a, Y_a, Z_a) \text{ by rotation } \Phi_a \text{ around } \overrightarrow{OX_a}$$

$$\begin{pmatrix} I_{af} \\ J_{af} \\ K_{af} \end{pmatrix} = \begin{pmatrix} \cos\gamma_a & 0 & \sin\gamma_a \\ 0 & 1 & 0 \\ -\sin\gamma_a & 0 & \cos\gamma_a \end{pmatrix} * \begin{pmatrix} 1 & 0 & 0 \\ 0 & \cos\Phi_a & -\sin\Phi_a \\ 0 & \sin\Phi_a & \cos\Phi_a \end{pmatrix} * \begin{pmatrix} I_a \\ J_a \\ Z_a \end{pmatrix}$$

$$\begin{pmatrix} I_{af} \\ J_{af} \\ K_{af} \end{pmatrix} = \begin{pmatrix} \cos\gamma_a & \sin\gamma_a * \sin\Phi_a & \sin\gamma_a * \cos\Phi_a \\ 0 & \cos\Phi_a & -\sin\Phi_a \\ -\sin\gamma_a & \cos\gamma_a * \sin\Phi_a & \cos\gamma_a * \cos\Phi_a \end{pmatrix} * \begin{pmatrix} I_a \\ J_a \\ Z_a \end{pmatrix}$$

We go from (O, X_0, Y_0, Z_0) to $(O, X_{af}, Y_{af}, Z_{af})$ by the following rotation:

$$(O, X_0, Y_0, Z_0) \rightarrow (O, X_{af}, Y_{af}, Z_0) = (O, X_{af}, Y_{af}, Z_{af})$$

by a rotation χ_a around $\overrightarrow{OZ_0}$.

However, it is also possible to perform the following five successive rotations:

$$(O, X_{af}, Y_{af}, Z_{af}) \rightarrow (O, X_h, Y_h, Z_{af}) \text{ by rotation } \delta \text{ around } \overrightarrow{OZ_{af}}$$

$$(O, X_h, Y_h, Z_{af}) \rightarrow (O, X_1, Y_h, Z_1') \text{ by rotation } \theta \text{ around } \overrightarrow{OY_h}$$

$$(O, X_1, Y_h, Z_1') \rightarrow (O, X_1, Y_1, Z_1) \text{ by rotation } \varphi \text{ around } \overrightarrow{OX_1}$$

$$(O, X_1, Y_1, Z_1) \rightarrow (O, X_2, Y_1, Z_a) \text{ by rotation } \alpha \text{ around } \overrightarrow{OY_1}$$

$$(O, X_2, Y_1, Z_a) \rightarrow (O, X_a, Y_a, Z_a) \text{ by rotation } \beta \text{ around } \overrightarrow{OZ_a}$$

$$\begin{pmatrix} I_{af} \\ J_{af} \\ K_{af} \end{pmatrix} = \begin{pmatrix} \cos\delta & -\sin\delta & 0 \\ \sin\delta & \cos\delta & 0 \\ 0 & 0 & 1 \end{pmatrix} * \begin{pmatrix} \cos\theta & 0 & \sin\theta \\ 0 & 1 & 0 \\ -\sin\theta & 0 & \cos\theta \end{pmatrix}$$
$$* \begin{pmatrix} 1 & 0 & 0 \\ 0 & \cos\Phi & -\sin\Phi \\ 0 & \sin\Phi & \cos\Phi \end{pmatrix} * \begin{pmatrix} \cos\alpha & 0 & -\sin\alpha \\ 0 & 1 & 0 \\ \sin\alpha & 0 & \cos\alpha \end{pmatrix}$$
$$* \begin{pmatrix} \cos\beta & -\sin\beta & 0 \\ \sin\beta & \cos\beta & 0 \\ 0 & 0 & 1 \end{pmatrix} * \begin{pmatrix} I_a \\ J_a \\ Z_a \end{pmatrix}$$

$$\begin{pmatrix} I_{af} \\ J_{af} \\ K_{af} \end{pmatrix} = \begin{pmatrix} \cos\delta & -\sin\delta & 0 \\ \sin\delta & \cos\delta & 0 \\ 0 & 0 & 1 \end{pmatrix} * \begin{pmatrix} \cos\theta & \sin\theta * \sin\Phi & \sin\theta * \cos\Phi \\ 0 & \cos\Phi & -\sin\Phi \\ -\sin\theta * \cos\theta & \sin\Phi & \cos\theta * \cos\Phi \end{pmatrix} * \begin{pmatrix} \cos\alpha * \cos\beta & -\cos\alpha * \sin\beta & -\sin\alpha \\ \sin\beta & \cos\beta & 0 \\ \sin\alpha * \cos\beta & -\sin\alpha * \sin\beta & \cos\alpha \end{pmatrix} * \begin{pmatrix} I_a \\ J_a \\ Z_a \end{pmatrix}$$

$$\begin{pmatrix} I_{af} \\ J_{af} \\ K_{af} \end{pmatrix} = \begin{pmatrix} \cos\delta & -\sin\delta & 0 \\ \sin\delta & \cos\delta & 0 \\ 0 & 0 & 1 \end{pmatrix} * \begin{pmatrix} M_{11} & M_{12} & M_{13} \\ M_{21} & M_{22} & M_{23} \\ M_{31} & M_{32} & M_{33} \end{pmatrix} * \begin{pmatrix} I_a \\ J_a \\ Z_a \end{pmatrix}$$

with:

$$M_{11} = (\cos\theta * \cos\alpha * \cos\beta) + (\sin\theta * \sin\Phi * \sin\beta) + (\sin\theta * \cos\Phi * \sin\alpha * \cos\beta)$$

$$M_{12} = (-(\cos\theta * \cos\alpha * \sin\beta) + (\sin\theta * \sin\Phi * \cos\beta) - (\sin\theta * \cos\Phi * \sin\alpha * \sin\beta))$$

$$M_{13} = (-(\cos\theta * \sin\alpha) + (\sin\theta * \cos\Phi * \cos\alpha))$$

$$M_{21} = ((\cos\Phi * \sin\beta) - (\sin\Phi * \sin\alpha * \cos\beta))$$

$$M_{22} = ((\cos\Phi * \cos\beta) + (\sin\Phi * \sin\alpha * \sin\beta)$$

$$M_{23} = -\sin\Phi * \cos\alpha$$

$$M_{31} = (-(\sin\theta * \cos\alpha * \cos\beta) + ((\cos\theta * \sin\Phi * \sin\beta) + (\cos\theta * \cos\Phi * \sin\alpha * \cos\beta)$$

$$M_{32} = ((\sin\theta * \cos\alpha * \sin\beta) + ((\cos\theta * \sin\Phi * \cos\beta) - ((\cos\theta * \cos\Phi * \sin\alpha * \sin\beta))$$

$$M_{33} = ((\sin\theta * \sin\alpha) + ((\cos\theta * \cos\Phi * \cos\alpha))$$

8.8.2.2. *Reduction of the equation*

From the identity of the matrix, we have:

$$\cos \gamma_a = \cos \delta ** M_{11} - \sin \delta * M_{21}$$

$$\cos \gamma_a = \cos \delta * [(\cos \theta * \cos \alpha * \cos \beta) + (\sin \theta * \sin \Phi * \sin \beta) + (\sin \theta * \cos \Phi * \sin \alpha * \cos \beta)] - \sin \delta * [((\cos \Phi * \sin \beta) - (\sin \Phi * \sin \alpha * \cos \beta))]$$

$$0 = \sin \delta * M_{11} + \cos \delta * M_{21}$$

$$0 = \sin \delta * [(\cos \theta * \cos \alpha * \cos \beta) + (\sin \theta * \sin \Phi * \sin \beta) + (\sin \theta * \cos \Phi * \sin \alpha * \cos \beta)] + \cos \delta * [((\cos \Phi * \sin \beta) - (\sin \Phi * \sin \alpha * \cos \beta))]$$

By posing either:

$$A = [(\cos \theta * \cos \alpha * \cos \beta) + (\sin \theta * \sin \Phi * \sin \beta) + (\sin \theta * \cos \Phi * \sin \alpha * \cos \beta)]$$

$$B = [((\cos \Phi * \sin \beta) - (\sin \Phi * \sin \alpha * \cos \beta))]$$

we obtain:

$$0 = \sin \delta * A + \cos \delta * B \Rightarrow \cos \delta = (-(\sin \delta * A)/B)$$

$$\cos \gamma_a = (\cos \delta * A) - (\sin \delta * B) = -((A^2 * \sin \delta)/B) - B * \sin \delta) = -(\sin \delta * (A^2 + B^2)/B)$$

$$\Rightarrow \sin \delta = \cos \gamma_a * (((-B)/(A^2 + B^2))$$

$$A^2 + B^2 = \cos \theta^2 * \cos \alpha^2 * \cos \beta^2 + \sin \theta^2 * \sin \phi^2 * \sin \beta^2 + \sin \theta^2 * \cos \phi^2 * \sin \alpha^2 * \cos \beta^2 + 2 * \cos \theta * \cos \alpha * \cos \beta * \sin \theta * \sin \phi * \sin \beta + 2 * \cos \theta * \cos \alpha * \cos \beta * \sin \theta * \cos \phi * \sin \alpha * \cos \beta + 2 * \sin \theta^2 * \sin \phi * \sin \beta * \cos \phi * \sin \alpha * \cos \beta + \cos \phi^2 * \sin \beta^2 + \sin \phi^2 * \sin \alpha^2 * \cos \beta^2 - 2 * \cos \phi * \sin \beta * \sin \phi * \sin \alpha * \cos \beta$$

$$A^2 + B^2 = \cos\theta^2 * \cos\alpha^2 * \cos\beta^2 + \sin\theta^2 * (\sin\phi^2 * \sin\beta^2 + \cos\phi^2 * \sin\alpha^2 * \cos\beta^2) + 2 * \cos\theta * \cos\alpha * \cos\beta * \sin\theta * (\sin\phi * \sin\beta + \cos\phi * \sin\alpha * \cos\beta) + \sin\theta^2 * (2 * \sin\phi * \sin\beta * \cos\phi * \sin\alpha * \cos\beta) + \cos\phi^2 * \sin\beta^2 + \sin\phi^2 * \sin\alpha^2 * \cos\beta^2 - 2 * \cos\phi * \sin\beta * \sin\phi * \sin\alpha * \cos\beta$$

$$A^2 + B^2 = \cos\theta^2 * \cos\alpha^2 * \cos\beta^2 + 2 * \cos\theta * \cos\alpha * \cos\beta * \sin\theta * (\sin\phi * \sin\beta + \cos\phi * \sin\alpha * \cos\beta) + \sin\theta^2 * (\sin\phi^2 * \sin\beta^2 + \cos\phi^2 * \sin\alpha^2 * \cos\beta^2) + \sin\theta^2 * (2 * \sin\phi * \sin\beta * \cos\phi * \sin\alpha * \cos\beta) + \cos\phi^2 * \sin\beta^2 + \sin\phi^2 * \sin\alpha^2 * \cos\beta^2 - 2 * \cos\phi * \sin\beta * \sin\phi * \sin\alpha * \cos\beta$$

$$A^2 + B^2 = \cos\theta^2 * \cos\alpha^2 * \cos\beta^2 + 2 * \cos\theta * \cos\alpha * \cos\beta * \sin\theta * (\sin\phi * \sin\beta + \cos\phi * \sin\alpha * \cos\beta) + \sin\theta^2 * (\sin\phi^2 * \sin\beta^2 + \cos\phi^2 * \sin\alpha^2 * \cos\beta^2 + 2 * \sin\phi * \sin\beta * \cos\phi * \sin\alpha * \cos\beta) + \cos\phi^2 * \sin\beta^2 + \sin\phi^2 * \sin\alpha^2 * \cos\beta^2 - 2 * \cos\phi * \sin\beta * \sin\phi * \sin\alpha * \cos\beta$$

$$A^2 + B^2 = \cos\theta^2 * \cos\alpha^2 * \cos\beta^2 + \sin\theta^2 * (\sin\phi * \sin\beta + \cos\phi * \sin\alpha * \cos\beta)^2 + 2 * \sin\theta * \cos\theta * \cos\alpha * \cos\beta\,(\sin\phi * \sin\beta + \cos\phi * \sin\alpha * \cos\beta) + \cos\phi^2 * \sin\beta^2 + \sin\phi^2 * \sin\alpha^2 * \cos\beta^2 - 2 * \cos\phi * \sin\beta * \sin\phi * \sin\alpha * \cos\beta$$

From the identity of the matrix, we also have:

$$\sin\gamma_a = ((\sin\theta * \cos\alpha * \cos\beta) - (\cos\theta * \sin\phi * \sin\beta) - (\cos\theta * \cos\phi * \sin\alpha * \cos\beta))$$

$$\sin{\gamma_a}^2 = \sin\theta^2 * \cos\alpha^2 * \cos\beta^2 + \cos\theta^2 * \sin\phi^2 * \sin\beta^2 + \cos\theta^2 * \cos\phi^2 * \sin\alpha^2 * \cos\beta^2 - 2 * \sin\theta * \cos\alpha * \cos\beta * \cos\theta * \sin\phi * \sin\beta - 2 * \sin\theta * \cos\alpha * \cos\beta * \cos\theta * \cos\phi * \sin\alpha * \cos\beta + 2 * \cos\theta * \sin\phi * \sin\beta * \cos\theta * \cos\phi * \sin\alpha * \cos\beta$$

$$\sin {\gamma_a}^2 = \sin \theta^2 * \cos \alpha^2 * \cos \beta^2 + \cos \theta^2 * \sin \phi^2 * \sin \beta^2 + \cos \theta^2 * \cos \phi^2 * \sin \alpha^2 * \cos \beta^2 + 2 * \cos \theta^2 * \sin \phi * \sin \beta * \cos \phi * \sin \alpha * \cos \beta - 2 * \sin \theta * \cos \alpha * \cos \beta * \cos \theta * (\sin \phi * \sin \beta + \cos \phi * \sin \alpha * \cos \beta)$$

$$\sin {\gamma_a}^2 = \sin \theta^2 * \cos \alpha^2 * \cos \beta^2 + \cos \theta^2 * (\sin \phi^2 * \sin \beta^2 + \cos \phi^2 * \sin \alpha^2 * \cos \beta^2 + 2 * \sin \phi * \sin \beta * \cos \phi * \sin \alpha * \cos \beta) - 2 * \sin \theta * \cos \alpha * \cos \beta * \cos \theta * (\sin \phi * \sin \beta + \cos \phi * \sin \alpha * \cos \beta)$$

$$\sin {\gamma_a}^2 = (1 - \cos \theta^2) * \cos \alpha^2 * \cos \beta^2 + \cos \theta^2 * (\sin \phi * \sin \beta + \cos \phi * \sin \alpha * \cos \beta)^2 - 2 * \sin \theta * \cos \alpha * \cos \beta * \cos \theta * (\sin \phi * \sin \beta + \cos \phi * \sin \alpha * \cos \beta)$$

$$\sin {\gamma_a}^2 = (1 - \cos \theta^2) * \cos \alpha^2 * \cos \beta^2 + (1 - \sin \theta^2) * (\sin \phi * \sin \beta + \cos \phi * \sin \alpha * \cos \beta)^2 - 2 * \sin \theta * \cos \alpha * \cos \beta * \cos \theta * (\sin \phi * \sin \beta + \cos \phi * \sin \alpha * \cos \beta)$$

$$\sin {\gamma_a}^2 = -(\cos \theta^2 * \cos \alpha^2 * \cos \beta^2) - (\sin \theta^2 * (\sin \phi * \sin \beta + \cos \phi * \sin \alpha * \cos \beta)^2) - 2 * \sin \theta * \cos \alpha * \cos \beta * \cos \theta * (\sin \phi * \sin \beta + \cos \phi * \sin \alpha * \cos \beta) + \cos \alpha^2 * \cos \beta^2 + \sin \phi^2 * \sin \beta^2 + \cos \phi^2 * \sin \alpha^2 * \cos \beta^2 + 2 * \sin \phi * \sin \beta * \cos \phi * \sin \alpha * \cos \beta$$

$$\sin {\gamma_a}^2 = -(\cos \theta^2 * \cos \alpha^2 * \cos \beta^2) - (\sin \theta^2 * (\sin \phi * \sin \beta + \cos \phi * \sin \alpha * \cos \beta)^2) - 2 * \sin \theta * \cos \alpha * \cos \beta * \cos \theta * (\sin \phi * \sin \beta + \cos \phi * \sin \alpha * \cos \beta) + (1 - \cos \phi^2) * \sin \beta^2 + (1 - \sin \phi^2) * \sin \alpha^2 * \cos \beta^2 + 2 * \sin \phi * \sin \beta * \cos \phi * \sin \alpha * \cos \beta + \cos \alpha^2 * \cos \beta^2$$

$$\begin{aligned}\sin\gamma_a{}^2 = -(&\cos\theta^2 * \cos\alpha^2 * \cos\beta^2)\\ &- (\sin\theta^2 * (\sin\phi * \sin\beta + \cos\phi * \sin\alpha * \cos\beta)^2)\\ &- 2 * \sin\theta * \cos\alpha * \cos\beta * \cos\theta\\ &* (\sin\phi * \sin\beta + \cos\phi * \sin\alpha * \cos\beta) + \sin\beta^2\\ &- \cos\phi^2 * \sin\beta^2 + \sin\alpha^2 * \cos\beta^2 - \sin\phi^2 * \sin\alpha^2\\ &* \cos\beta^2 + 2 * \sin\phi * \sin\beta * \cos\phi * \sin\alpha * \cos\beta\\ &+ \cos\alpha^2 * \cos\beta^2\end{aligned}$$

$$\begin{aligned}\sin\gamma_a{}^2 = -(&\cos\theta^2 * \cos\alpha^2 * \cos\beta^2 - \sin\theta^2\\ &* (\sin\phi * \sin\beta + \cos\phi * \sin\alpha * \cos\beta)^2 + 2 * \sin\theta\\ &* \cos\alpha * \cos\beta * \cos\theta\\ &* (\sin\phi * \sin\beta + \cos\phi * \sin\alpha * \cos\beta) + \cos\phi^2\\ &* \sin\beta^2 + \sin\phi^2 * \sin\alpha^2 * \cos\beta^2 - 2 * \sin\phi * \sin\beta\\ &* \cos\phi * \sin\alpha * \cos\beta) + \sin\beta^2 + \sin\alpha^2 * \cos\beta^2\\ &+ \cos\alpha^2 * \cos\beta^2\end{aligned}$$

We can conclude that:

$$\sin\gamma_a{}^2 = -(A^2 + B^2) + \sin\beta^2 + \sin\alpha^2 * \cos\beta^2 + \cos\alpha^2 * \cos\beta^2$$

from which we obtain:

$$\begin{aligned}(A^2 + B^2) &= -\sin\gamma_a{}^2 + \sin\beta^2 + \sin\alpha^2 * \cos\beta^2 + \cos\alpha^2 * \cos\beta^2\\ &= -\sin\gamma_a{}^2 + \sin\beta^2 + \cos\beta^2\,(\sin\alpha^2 + \cos\alpha^2)\\ &= -\sin\gamma_a{}^2 + \sin\beta^2 + \cos\beta^2 = 1 - \sin\gamma_a{}^2\end{aligned}$$

$$(A^2 + B^2) = 1 - \sin\gamma_a{}^2 = \cos\gamma_a{}^2$$

By taking the equation:

$$\sin\delta = \cos\gamma_a * (((-B)/(A^2 + B^2))$$

we obtain:

$$\begin{aligned}\sin\delta = \cos\gamma_a * \big(((-B)/(A^2 + B^2))\big) &= (-B) * (\cos\gamma_a/\cos\gamma_a{}^2)\\ &= ((-B)/\cos\gamma_a)\end{aligned}$$

Let:

$$\sin\delta = \frac{(\cos\phi * \sin\beta) - (\sin\phi * \sin\alpha * \cos\beta)}{\cos\gamma_a}$$

We can conclude that:

$$\delta = \text{Arcsin}(\sin\delta)$$

and:

$$\Psi_{ac} = \chi_{ac} + \delta$$

$$\Psi_a = \chi_a + \delta$$

8.8.3. *Determination of θ*

If we take the equations of the previous section, we obtain:

$$0 = \sin\delta * A + \cos\delta * B$$

$$\cos\gamma_a = (\cos\delta * A) - (\sin\delta * B)$$

which give:

$$0 = \sin\delta * \sin\delta * A + \sin\delta * \cos\delta * B$$

$$\cos\gamma_a * \cos\delta = (\cos\delta * \cos\delta * A) - (\sin\delta * \cos\delta * B)$$

$$0 = \sin\delta^2 * A + \sin\delta * \cos\delta * B$$

$$\cos\gamma_a * \cos\delta = (\cos\delta^2 * A) - (\sin\delta * \cos\delta * B)$$

$$\Rightarrow 0 + \cos\gamma_a * \cos\delta = \sin\delta^2 * A + \sin\delta * \cos\delta * B + \cos\delta^2 * A - \sin\delta * \cos\delta * B$$

$$\Rightarrow \cos\gamma_a * \cos\delta = \sin\delta^2 * A + \cos\delta^2 * A$$

$$\Rightarrow A = \cos\gamma_a * \cos\delta$$

knowing that:

$$A = [(\cos\theta * \cos\alpha * \cos\beta) + (\sin\theta * \sin\Phi * \sin\beta) + (\sin\theta * \cos\Phi * \sin\alpha * \cos\beta)]$$

$$\sin\gamma_a = ((\sin\theta * \cos\alpha * \cos\beta) - (\cos\theta * \sin\phi * \sin\beta) - (\cos\theta * \cos\phi * \sin\alpha * \cos\beta))$$

By letting:

$$C = \cos\alpha * \cos\beta$$

$$D = \sin\phi * \sin\beta + \cos\phi * \sin\alpha * \cos\beta$$

we can then write:

$$\frac{C * \sin\theta - D * \cos\theta}{C * \cos\theta + D * \sin\theta} = \frac{\sin\gamma_a}{\cos\gamma_a * \cos\delta} = \frac{C * \tan\theta - D}{C + D * \tan\theta}$$

$$C * \tan\theta * \cos\gamma_a * \cos\delta - D * \cos\gamma_a * \cos\delta = C * \sin\gamma_a + D * \tan\theta * \sin\gamma_a$$

$$C * \tan\theta * \cos\gamma_a * \cos\delta - D * \tan\theta * \sin\gamma_a = C * \sin\gamma_a + D * \cos\gamma_a * \cos\delta$$

$$\tan\theta = \frac{C * \sin\gamma_a + D * \cos\gamma_a * \cos\delta}{C * \cos\gamma_a * \cos\delta - D * \sin\gamma_a}$$

$$\tan\theta = \frac{\cos\alpha * \cos\beta * \sin\gamma_a + (\sin\phi * \sin\beta + \cos\phi * \sin\alpha * \cos\beta) * \cos\gamma_a * \cos\delta}{\cos\alpha * \cos\beta * \cos\gamma_a * \cos\delta - (\sin\phi * \sin\beta + \cos\phi * \sin\alpha * \cos\beta) * \sin\gamma_a}$$

8.9. Determination of load factors (n_{x1}, n_{y1}, n_{z1}) in the aircraft trihedron

To pass from the trihedron (O, X_a, Y_{af}, Z_r) to the aircraft trihedron (O, X_1, Y_1, Z_1), we carry out three successive rotations:

$(O, X_a, Y_{af}, Z_r) \rightarrow (O, X_a, Y_a, Z_a)$ by rotation ϕ_a around $\overrightarrow{O\,X_a}$

$(O, X_a, Y_a, Z_a) \rightarrow (O, X_{a1}, Y_1, Z_a)$ by rotation β around $\overrightarrow{O\,Z_a}$

$(O, X_{a1}, Y_1, Z_a) \rightarrow (O, X_1, Y_1, Z_1)$ by rotation α around $\overrightarrow{O\,Y_1}$

$$\begin{pmatrix} I_a \\ J_{af} \\ K_r \end{pmatrix} = \begin{pmatrix} 1 & 0 & 0 \\ 0 & \cos\phi_a & -\sin\phi_a \\ 0 & \sin\phi_a & \cos\phi_a \end{pmatrix} * \begin{pmatrix} \cos\beta & \sin\beta & 0 \\ -\sin\beta & \cos\beta & 0 \\ 0 & 0 & 1 \end{pmatrix} * \begin{pmatrix} \cos\alpha & 0 & \sin\alpha \\ 0 & 1 & 0 \\ -\sin\alpha & 0 & \cos\alpha \end{pmatrix} * \begin{pmatrix} I_1 \\ J_1 \\ Z_1 \end{pmatrix}$$

$$\begin{pmatrix} I_a \\ J_{af} \\ K_r \end{pmatrix} = \begin{pmatrix} \cos\beta & \sin\beta & 0 \\ -\cos\phi_a * \sin\beta & \cos\phi_a * \cos\beta & -\sin\phi_a \\ -\sin\phi_a * \sin\beta & \sin\phi_a * \cos\beta & \cos\phi_a \end{pmatrix} * \begin{pmatrix} \cos\alpha & 0 & \sin\alpha \\ 0 & 1 & 0 \\ -\sin\alpha & 0 & \cos\alpha \end{pmatrix} * \begin{pmatrix} I_1 \\ J_1 \\ Z_1 \end{pmatrix}$$

$$\begin{pmatrix} I_a \\ J_{af} \\ K_r \end{pmatrix} = \begin{pmatrix} M_{11} & M_{12} & M_{13} \\ M_{21} & M_{22} & M_{23} \\ M_{31} & M_{32} & M_{33} \end{pmatrix} * \begin{pmatrix} I_1 \\ J_1 \\ Z_1 \end{pmatrix}$$

The inverse matrix, which makes it possible to pass from the aircraft trihedron to the reference trihedron, is symmetrical compared to the diagonal:

$$\begin{pmatrix} I_1 \\ J_1 \\ Z_1 \end{pmatrix} = \text{Matrix}^{-1} * \begin{pmatrix} I_a \\ J_{af} \\ K_r \end{pmatrix}$$

$$\text{Matrix}^{-1} = \begin{pmatrix} M_{11} & M_{21} & M_{31} \\ M_{12} & M_{22} & M_{32} \\ M_{13} & M_{23} & M_{33} \end{pmatrix}$$

with:

$$M_{11} = \cos\beta * \cos\alpha$$

$$M_{12} = \sin\beta$$

$$M_{13} = \cos\beta * \sin\alpha$$

$$M_{21} = -\cos\phi_a * \sin\beta * \cos\alpha + \sin\phi_a * \sin\alpha$$

$$M_{22} = \cos\phi_a * \cos\beta$$

$$M_{23} = -\cos\phi_a * \sin\beta * \sin\alpha - \sin\phi_a * \cos\alpha$$

$$M_{31} = -\sin\phi_a * \sin\beta * \cos\alpha - \cos\phi_a * \sin\alpha$$

$$M_{32} = \sin\phi_a * \cos\beta$$

$$M_{33} = -\sin\phi_a * \sin\beta * \sin\alpha + \cos\phi_a * \cos\alpha$$

However, according to section 8.8.2.1 and the matrix identities, it is possible to write:

$$\cos\gamma_a * \sin\phi_a = \sin\theta * \cos\alpha * \sin\beta + \cos\theta * (\sin\phi * \cos\beta - \cos\phi * \sin\alpha * \sin\beta)$$

$$\cos\gamma_a * \cos\phi_a = \sin\theta * \sin\alpha + \cos\theta * \cos\phi * \cos\alpha$$

from which we obtain:

$$\sin\phi_a = \frac{\sin\theta * \cos\alpha * \sin\beta + \cos\theta * (\sin\phi * \cos\beta - \cos\phi * \sin\alpha * \sin\beta)}{\cos\gamma_a}$$

$$\cos\phi_a = \frac{\sin\theta * \sin\alpha + \cos\theta * \cos\phi * \cos\alpha}{\cos\gamma_a}$$

Therefore, the load factors are given by the Matrix^{-1} defined above:

$$\begin{pmatrix} n_{x1} \\ n_{y1} \\ n_{z1} \end{pmatrix} = \text{Matrix}^{-1} * \begin{pmatrix} n_{ia} \\ n_{jaf} \\ n_{kr} \end{pmatrix}$$

8.10. Determination of C_Y and C_Z

We have:

$$C_Y = \frac{m * n_{y1} * g}{Q * S} = \frac{I * C_{Yr} * r_1}{V_a}$$

$$C_Z = \frac{m * n_{z1} * g}{Q * S} = \frac{I * C_{Zq} * q_1}{V_a}$$

When n_{y1} and n_{z1} come from telemetry measurements, these values can be used to calculate C_Y and C_Z (a single iteration).

If we do not have the values p_1, q_1 and r_1, we must take q_1 and r_1 to be equal to 0 to start the calculation.

We must be careful, however: when these data come from telemetry values, we must know the measurement conditions. Indeed, if the recordings n_{y1} and n_{z1} are given at zero roll, we note them as $n_{y1\phi}$ and $n_{z1\phi}$:

$$n_{y1} = n_{y1\phi} * \cos\phi + n_{z1\phi} * \sin\phi$$

$$n_{z1} = -(n_{y1\phi} * \sin\phi) + n_{z1\phi} * \cos\phi$$

8.11. Determination of the total incidence

The total incidence is given by:

$$\alpha_t = \sqrt{\alpha^2 + \beta^2}$$

8.12. Determination of the longitudinal attitude compared to the local horizontal plane

To move from the frame (O, X_a, Y_{af}, Z_r) to the local geographic frame $(O, X_{VL}, Y_{VL}, Z_{VL})$, we use the following two sets of rotations.

The first series of rotations is as follows:

$(O, X_a, Y_{af}, Z_r) \rightarrow (O, X_a, Y_k, Z_k)$ by rotation ρ around $\overrightarrow{O X_a}$

$(O, X_a, Y_k, Z_k) \rightarrow (O, X_k, Y_k, Z_{VL})$ by rotation θ_{VL} around $\overrightarrow{O Y_k}$

$(O, X_k, Y_k, Z_{VL}) \rightarrow (O, X_{VL}, Y_{VL}, Z_{VL})$ by rotation Ψ_{VL} around $\overrightarrow{O Z_{VL}}$

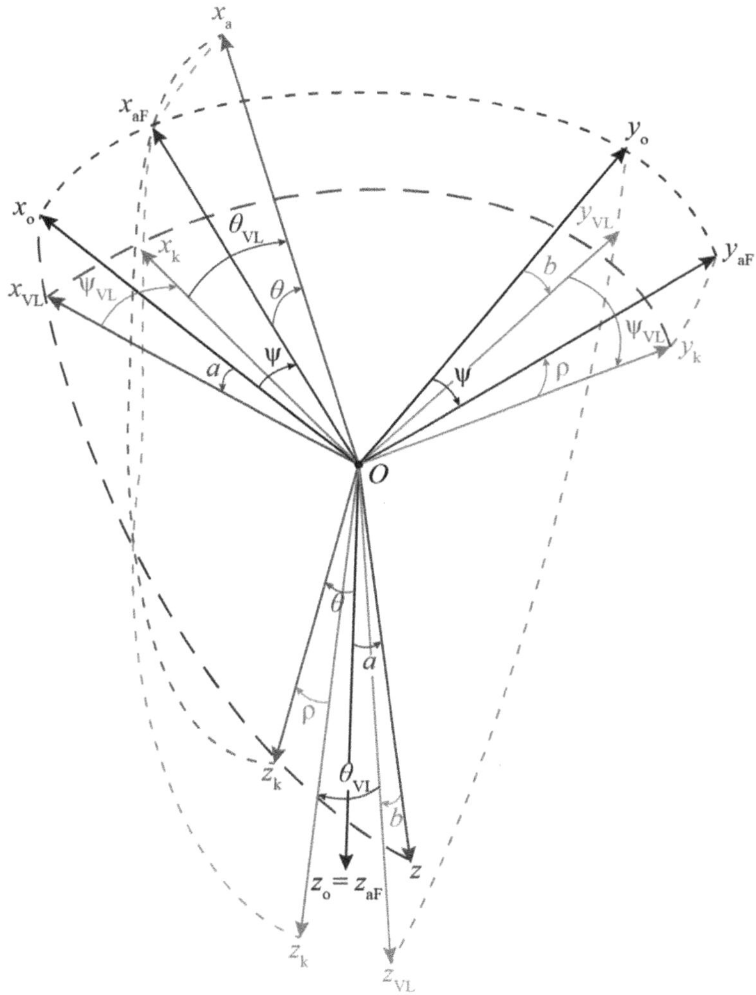

Figure 8.3. *Trihedron linked to local vertical. For a color version of this figure, see www.iste.co.uk/louis/flight.zip*

The second series of rotations is as follows:

$(O, X_a, Y_{af}, Z_r) \rightarrow (O, X_{af}, Y_{af}, Z_0)$ by rotation θ around $\overrightarrow{O\,Y_{af}}$

$(O, X_{af}, Y_{af}, Z_0) \rightarrow (O, X_0, Y_0, Z_0)$ by rotation Ψ around $\overrightarrow{O\,Z_0}$

$(O, X_0, Y_0, Z_0) \rightarrow (O, X_{VL}, Y_0, Z)$ by rotation a around $\overrightarrow{OY_0}$

$(O, X_{VL}, Y_0, Z) \rightarrow (O, X_{VL}, Y_{VL}, Z_{VL})$ by rotation b around $\overrightarrow{O\,X_{VL}}$

The resolution of the first series of rotations gives:

$$\begin{pmatrix} I_a \\ J_{af} \\ K_r \end{pmatrix} = \begin{pmatrix} 1 & 0 & 0 \\ 0 & \cos\rho & \sin\rho \\ 0 & -\sin\rho & \cos\rho \end{pmatrix} * \begin{pmatrix} \cos\theta_{VL} & 0 & -\sin\theta_{VL} \\ 0 & 1 & 0 \\ \sin\theta_{VL} & 0 & \cos\theta_{VL} \end{pmatrix} * \begin{pmatrix} \cos\Psi_{VL} & \sin\Psi_{VL} & 0 \\ -\sin\Psi_{VL} & \cos\Psi_{VL} & 0 \\ 0 & 0 & 1 \end{pmatrix} * \begin{pmatrix} I_1 \\ J_1 \\ Z_1 \end{pmatrix}$$

$$\begin{pmatrix} I_a \\ J_{af} \\ K_r \end{pmatrix} = \begin{pmatrix} \cos\theta_{VL} & 0 & -\sin\theta_{VL} \\ \sin\rho * \sin\theta_{VL} & \cos\rho & \sin\rho * \cos\theta_{VL} \\ \cos\rho * \sin\theta_{VL} & -\sin\rho & \cos\rho * \cos\theta_{VL} \end{pmatrix} * \begin{pmatrix} \cos\Psi_{VL} & \sin\Psi_{VL} & 0 \\ -\sin\Psi_{VL} & \cos\Psi_{VL} & 0 \\ 0 & 0 & 1 \end{pmatrix} * \begin{pmatrix} I_1 \\ J_1 \\ Z_1 \end{pmatrix}$$

$$\begin{pmatrix} I_a \\ J_{af} \\ K_r \end{pmatrix} = \begin{pmatrix} M_{11} & M_{12} & M_{13} \\ M_{21} & M_{22} & M_{23} \\ M_{31} & M_{32} & M_{33} \end{pmatrix} * \begin{pmatrix} I_1 \\ J_1 \\ Z_1 \end{pmatrix}$$

The inverse matrix, which makes it possible to pass from the aircraft trihedron to the reference trihedron, is symmetrical compared to the diagonal:

$$\begin{pmatrix} I_1 \\ J_1 \\ Z_1 \end{pmatrix} = \text{Matrix}^{-1} * \begin{pmatrix} I_a \\ J_{af} \\ K_r \end{pmatrix}$$

$$\text{Matrix}^{-1} = \begin{pmatrix} M_{11} & M_{21} & M_{31} \\ M_{12} & M_{22} & M_{32} \\ M_{13} & M_{23} & M_{33} \end{pmatrix}$$

with:

$$M_{11} = \cos\theta_{VL} * \cos\Psi_{VL}$$

$$M_{12} = \cos\theta_{VL} * \sin\Psi_{VL}$$

$$M_{13} = -\sin\theta_{VL}$$

$$M_{21} = \sin\rho * \sin\theta_{VL} * \cos\Psi_{VL} - \cos\rho * \sin\Psi_{VL}$$

$$M_{22} = \sin\rho * \sin\theta_{VL} * \sin\Psi_{VL} + \cos\rho * \cos\Psi_{VL}$$

$$M_{23} = \sin\rho * \cos\theta_{VL}$$

$$M_{31} = \cos\rho * \sin\theta_{VL} * \cos\Psi_{VL} + \sin\rho * \sin\Psi_{VL}$$

$$M_{32} = \cos\rho * \sin\theta_{VL} * \sin\Psi_{VL} - \sin\rho * \cos\Psi_{VL}$$

$$M_{33} = \cos\rho * \cos\theta_{VL}$$

The resolution of the second series of rotations gives:

$$\begin{pmatrix} I_a \\ J_{af} \\ K_r \end{pmatrix} = \begin{pmatrix} \cos\theta & 0 & -\sin\theta \\ 0 & 1 & 0 \\ \sin\theta & 0 & \cos\theta \end{pmatrix} * \begin{pmatrix} \cos\Psi & \sin\Psi & 0 \\ -\sin\Psi & \cos\Psi & 0 \\ 0 & 0 & 1 \end{pmatrix} * \begin{pmatrix} \cos a & 0 & -\sin a \\ 0 & 1 & 0 \\ \sin a & 0 & \cos a \end{pmatrix} * \begin{pmatrix} 1 & 0 & 0 \\ 0 & \cos b & -\sin b \\ 0 & \sin b & \cos b \end{pmatrix} * \begin{pmatrix} I_1 \\ J_1 \\ Z_1 \end{pmatrix}$$

$$\begin{pmatrix} I_a \\ J_{af} \\ K_r \end{pmatrix} = \begin{pmatrix} \cos\theta * \cos\Psi & \cos\theta * \sin\Psi & -\sin\theta \\ -\sin\Psi & \cos\Psi & 0 \\ \sin\theta * \cos\Psi & \sin\theta * \sin\Psi & \cos\theta \end{pmatrix} * \begin{pmatrix} \cos a & -\sin a * \sin b & -\sin a * \cos b \\ 0 & \cos b & -\sin b \\ \sin a & \cos a * \sin b & \cos a * \cos b \end{pmatrix} * \begin{pmatrix} I_1 \\ J_1 \\ Z_1 \end{pmatrix}$$

According to the small angle assumptions, we can assimilate $\cos a$ at 1, $\cos b$ at 1, $\sin a$ at a and $\sin b$ at b.

Therefore, the above matrix can be written:

$$\begin{pmatrix} I_a \\ J_{af} \\ K_r \end{pmatrix} = \begin{pmatrix} \cos\theta * \cos\Psi & \cos\theta * \sin\Psi & -\sin\theta \\ -\sin\Psi & \cos\Psi & 0 \\ \sin\theta * \cos\Psi & \sin\theta * \sin\Psi & \cos\theta \end{pmatrix} * \begin{pmatrix} 1 & 0 & -a \\ 0 & 1 & -b \\ a & b & 1 \end{pmatrix} * \begin{pmatrix} I_1 \\ J_1 \\ Z_1 \end{pmatrix}$$

$$\begin{pmatrix} I_a \\ J_{af} \\ K_r \end{pmatrix} = \begin{pmatrix} M_{11} & M_{12} & M_{13} \\ M_{21} & M_{22} & M_{23} \\ M_{31} & M_{32} & M_{33} \end{pmatrix} * \begin{pmatrix} I_1 \\ J_1 \\ Z_1 \end{pmatrix}$$

with:

$$M_{11} = \cos\theta * \cos\Psi - a * \sin\theta$$

$$M_{12} = \cos\theta * \sin\Psi - b * \sin\theta$$

$$M_{13} = -a * \cos\theta * \cos\Psi - b * \cos\theta * \sin\Psi - \sin\theta$$

$$M_{21} = -\sin\Psi$$

$$M_{22} = \cos\Psi$$

$$M_{23} = a * \sin\Psi - b * \cos\Psi$$

$$M_{31} = \sin\theta * \cos\Psi + a * \cos\theta$$

$$M_{32} = \sin\theta * \sin\Psi + b * \cos\theta$$

$$M_{33} = -a * \sin\theta * \cos\Psi - b * \sin\theta * \sin\Psi + \cos\theta$$

Following the identity of the matrices, we can write:

$$\sin\theta_{VL} = a * \cos\theta * \cos\Psi + b * \cos\theta * \sin\Psi + \sin\theta$$

$$\theta_{VL} = \text{Arcsin}(a * \cos\theta * \cos\Psi + b * \cos\theta * \sin\Psi + \sin\theta)$$

with:

$$a = \frac{x_0}{R} \text{ and } b = \frac{y_0}{R}$$

8.13. Determination of drag coefficients

The drag coefficient is given by the following equation:

$$C_x = -\frac{FP - m * x_1 * g}{Q * S}$$

The induced drag coefficient is given by:

$$\Delta C_{x1} = f(M, i_T, (|\eta| + |\zeta|)$$

It is obtained from tables by three-dimensional interpolation.

The rudder drag coefficient is given by:

$$\Delta Cx2 = f(M, (\Sigma|\sigma|)$$

It is obtained from tables by two-dimensional interpolation, with:

$$\Sigma|\sigma| = |\sigma 1| + |\sigma 2| + |\sigma 3| + |\sigma 4|$$

and:

$$\sigma_1 = \sigma_2 = \sigma_3 = \sigma_4 = \frac{1 * (\eta + \zeta) + \xi}{2^{0,5}}$$

The minimum drag coefficient is given by:

$$C_{x0} = C_x - \Delta C_{x1} - \Delta C_{x2}$$

8.14. Determination of the pushing force

Knowing the theoretical drag coefficient, it is possible to reconstruct the real thrust of the thrusters:

$$FP = (Q * S * C_{xth}) - (m * x_1 * g)$$

where:

$$C_{xth} = C_{x0th} + \Delta C_{x1} + \Delta C_{x2}$$

9

Quaternion Methods

9.1. Goal

The problem posed in this chapter is that of simplifying the calculation of the change of axes by not making use of Euler angles, thereby avoiding the calculation of trigonometric lines and at the same time excluding any angular indeterminacy (a problem which arises from using Euler angles).

A method for changing axes that uses only rational functions can save calculation time (in the case of ground resources) or lead to a reduction in the size of the calculator (in the case of on-board equipment).

Euler angles are widely used in mechanics and astronomy. They are useful when explicit knowledge of a mobile around its center of gravity is required. In addition, they correspond to degrees of freedom materialized physically.

Olinde-Rodrigues's formulas make it possible to simplify and reduce the number of axis change calculations when explicit knowledge of angles Ψ, θ and φ is not required. They use the direction of the instantaneous axis of rotation (two parameters) and the angle of rotation V around this axis as three independent parameters.

9.2. Reminder of the axis change formulas using Euler angles

The most general change of axes corresponds to a certain displacement. It consists of a translation (movement of the center of gravity) and of a rotation around an axis passing through the center of gravity (movement around the center of gravity, i.e. around the instantaneous axis of rotation).

If we are only interested in the rotation around the center of gravity, and thus if Ω is the instantaneous speed vector of rotation:

$$\vec{\Omega} = p * \vec{i} + q * \vec{j} + r * \vec{k}$$

$\vec{i}, \vec{j}$ and $\vec{k}$ denote the unit vectors of the axes ox, oy and oz deduced from the fixed trihedron (O, X, Y, Z) from a rotation of an angle V around the instantaneous axis of rotation (Δ) passing through O.

$\vec{I}, \vec{J}$ and $\vec{K}$ designate the unit vectors of axes OX, OY and OZ.

By proceeding in order to the three rotations corresponding to the angles Ψ, θ and φ, we obtain the results explained in Chapter 5, section 5.5.1.

The integration of Ψ, θ and ϕ, knowing p, q and r and the Euler angles at time T, allows us to deduce the values of the Euler angles at the instant $T + \Delta T$.

The use of computer programs involves trigonometric line subroutines and requires solving the problem of determining the angles when $\theta = 90°$.

9.3. Olinde-Rodrigues's formulas: definition of quaternions

Olinde-Rodrigues's formulas express the nine director cosines as a rational function of three parameters directly related to the amplitude V of the rotation and to the director cosines of its axis (α, β, γ).

Let V therefore be the amplitude of the rotation, bringing the point M to M', and let H be the foot of the perpendicular lowered by ω on MM'.

The plane (ωMM') is normal to the axis of rotation (Δ) of unit vector u.

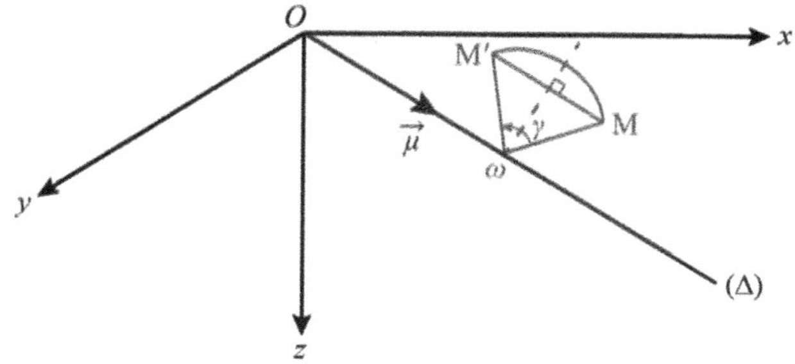

Figure 9.1. *Rotation axis Δ. For a color version of this figure, see www.iste.co.uk/louis/flight.zip*

Let $\overrightarrow{OM} = x * \vec{I} + y * \vec{J} + z * \vec{K}$:

$$OM' = x' * I + y' * J + z' * K = x * i + y * j + z * k$$

By letting:

$$\vec{u}_1 = \vec{u} * \tan\frac{V}{2}$$

we obtain:

$$\overrightarrow{HM'} = \overrightarrow{OM'} - \overrightarrow{OH} = \frac{\overrightarrow{OM'} - \overrightarrow{OM}}{2}$$

Thus:

$$\overrightarrow{HM'} = \frac{x - x'}{2} * \vec{I} + \frac{y - y'}{2} * \vec{J} + \frac{z - z'}{2} * \vec{K}$$

with:

$$\vec{u}_1 \Lambda \overrightarrow{\omega H} = \overrightarrow{HM'}(1)$$

since:

$$|\omega H| = \frac{\overrightarrow{HM'}}{\tan\frac{V}{2}}$$

$$\vec{u}_1 = \tan\frac{V}{2} * (\alpha * \vec{I} + \beta * \vec{J} + \gamma * \vec{K})$$

$$\vec{u}_1 \Lambda \overrightarrow{\omega H} = \vec{u}_1 \Lambda (\overrightarrow{OH} - \overrightarrow{O\omega}) = \vec{u}_1 \Lambda \overrightarrow{OH} = \vec{u}_1 \Lambda \frac{(\overrightarrow{OM} + \overrightarrow{OM'})}{2} \qquad [9.1]$$

from which we obtain:

$$\vec{u}_1 \Lambda \overrightarrow{\omega H} = \tan\frac{V}{2} * \left[\left(\beta * \frac{z+z'}{2} - \gamma * \frac{y+y'}{2}\right) * \vec{I} + \left(\gamma * \frac{x+x'}{2} - \alpha * \frac{z+z'}{2}\right) * \vec{J} + \left(\alpha * \frac{y+y'}{2} - \beta \frac{x+x'}{2}\right) * \vec{K}\right]$$

On the other hand:

$$\overrightarrow{HM'} = \frac{x-x'}{2} * \vec{I} + \frac{y-y'}{2} * \vec{J} + \frac{z-z'}{2} * \vec{K}$$

According to equation [9.1], we can write:

$$\tan\frac{V}{2} * \left(\beta * \frac{z+z'}{2} - \gamma * \frac{y+y'}{2}\right) = \frac{x'-x}{2}$$

$$\tan\frac{V}{2} * \left(\gamma * \frac{x+x'}{2} - \alpha * \frac{z+z'}{2}\right) = \frac{y'-y}{2}$$

$$\tan\frac{V}{2} * \left(\alpha * \frac{y+y'}{2} - \beta * \frac{x+x'}{2}\right) = \frac{z'-z}{2}$$

Hence, the linear relations linking x', y' and z' to x, y and z are:

$$x' + \gamma * \tan\frac{V}{2} * y' - \beta * \tan\frac{V}{2} * z' = x - \gamma * \tan\frac{V}{2} * y + \beta * \tan\frac{V}{2} * z$$

$$y' - \gamma * \tan\frac{V}{2} * x' + \alpha * \tan\frac{V}{2} * z' = y + \gamma * \tan\frac{V}{2} * x - \alpha * \tan\frac{V}{2} * z$$

$$z' + \beta * \tan\frac{V}{2} * x' - \alpha * \tan\frac{V}{2} * y' = z - \beta * \tan\frac{V}{2} * x + \alpha * \tan\frac{V}{2} * y$$

The resolution of the system gives:

$$\Delta = \begin{vmatrix} 1 & \gamma * \tan\frac{V}{2} & -\beta * \tan\frac{V}{2} \\ -\gamma * \tan\frac{V}{2} & 1 & \alpha * \tan\frac{V}{2} \\ \beta * \tan\frac{V}{2} & -\alpha * \tan\frac{V}{2} & 1 \end{vmatrix} = 1 + (\alpha^2 + \beta^2 + \gamma^2) * \tan\frac{V}{2}$$

from which we obtain:

$$\Delta = 1 + (\alpha^2 + \beta^2 + \gamma^2) * \tan\frac{V}{2} = 1 + \tan^2\frac{V}{2} = \frac{1}{\cos^2\frac{V}{2}} \qquad [9.2]$$

$$\Delta * x' = x * \left[1 + (\alpha^2 + \beta^2 + \gamma^2) * \tan^2\frac{V}{2}\right] + 2 * \left(\alpha * \beta * \tan\frac{V}{2} - \gamma\right) * \tan\frac{V}{2} * y + 2 * \left(\beta + \alpha * \gamma * \tan\frac{V}{2}\right) * \tan\frac{V}{2} * z$$

$$\Delta * y' = y * \left[1 + (-\alpha^2 + \beta^2 - \gamma^2) * \tan^2\frac{V}{2}\right] + 2 * \left(\alpha * \beta * \tan\frac{V}{2} + \gamma\right) * \tan\frac{V}{2} * x + 2 * \left(\beta * \gamma * \tan\frac{V}{2} - \gamma\right) * \tan\frac{V}{2} * z$$

$$\Delta * z' = 2 * \left(\alpha * \gamma * \tan\frac{V}{2} - \beta\right) * \tan\frac{V}{2} * x + 2 * \left(\alpha + \beta * \gamma * \tan\frac{V}{2}\right) * \tan\frac{V}{2} * y + 2 * \left[1 + (-\alpha^2 - \beta^2 + \gamma^2) * \tan^2\frac{V}{2}\right] * z$$

By letting:

$$\alpha * \tan\frac{V}{2} = l$$

$$\beta * \tan\frac{V}{2} = m\gamma * \tan\frac{V}{2} = n$$

we obtain:

$$\Delta * x' = x * [1 + l^2 - m^2 - n^2] + 2 * (m * l - n) * y + 2 * (m + l * n) * z$$

$$\Delta * y' = y * [1 - l^2 + m^2 - n^2] + 2 * (l * m + n) * x + 2 * (m * n - l) * z$$

$$\Delta * z' = z * [1 - l^2 - m^2 + n^2] + 2 * (l * n - m) * x + 2 * (l + m * n) * y$$

Thus:

$$x' = \overrightarrow{OM'} * \vec{I} = x * \vec{i} * \vec{I} + y * \vec{j} * \vec{I} + z * \vec{k} * \vec{I}$$

$$y' = \overrightarrow{OM'} * \vec{J} = x * \vec{i} * \vec{J} + y * \vec{j} * \vec{J} + z * \vec{k} * \vec{J}$$

$$z' = \overrightarrow{OM'} * \vec{K} = x * \vec{i} * \vec{K} + y * \vec{j} * \vec{K} + z * \vec{k} * \vec{K}$$

The nine directing cosines of the vectors $(\vec{I}, \vec{J}, \vec{K})$ in the trihedron $(\vec{i}, \vec{j}, \vec{k})$ are given by the following equations:

$$\vec{I} = \frac{[(1 + l^2 - m^2 - n^2) * \vec{i} + 2 * (m * l - n) * \vec{j} + 2 * (m + l * n) * \vec{k}]}{\Delta}$$

$$\vec{J} = \frac{[2 * (l * m + n) * \vec{i} + (1 - l^2 + m^2 - n^2) * \vec{j} + 2 * (m * n - l) * \vec{k}]}{\Delta}$$

$$\vec{K} = \frac{[2 * (l * n - m) * \vec{i} + 2 * (l + m * n) * \vec{j} + [1 - l^2 - m^2 + n^2] * \vec{k}]}{\Delta} \quad [9.3]$$

If we compare them to the formulas obtained for the directing cosines of $(\vec{I}, \vec{J}, \vec{K})$ with the Euler angle method (see section 4.5.1), we find:

$$\cos^2(V/2) * (1 + l^2 - m^2 - n^2) = \cos\Psi * \cos\theta = \vec{I} * \vec{i}$$

$$2 * \cos^2(V/2) * (l * m - n) = ((-\sin\Psi * \cos\phi) + (\cos\Psi * \sin\theta * \sin\phi)) = \vec{I} * \vec{j} \quad [9.4]$$

$$2 * \cos^2(V/2) * (l * n + m) = ((\sin\Psi * \sin\phi) + (\cos\Psi * \sin\theta * \cos\phi)) = \vec{I} * \vec{k}$$

$$2 * \cos^2(V/2) * (l * m + n) = \sin\Psi * \cos\theta = \vec{J} * \vec{i}$$

$$\cos^2(V/2) * (1 - l^2 + m^2 - n^2) \cos = ((\cos\Psi * \cos\phi) + (\sin\Psi * \sin\theta * \sin\phi)) = \vec{J} * \vec{j} \quad [9.5]$$

$$2 * \cos^2\left(\frac{V}{2}\right) * (m * n - l) = ((-\cos\Psi * \sin\phi) + (\sin\Psi * \sin\theta * \cos\phi)) = \vec{J} * \vec{k}$$

$$2 * \cos^2(V/2) * (l * m - n) = (-\sin\theta) = \vec{K} * \vec{i}$$

$$2 * \cos^2(V/2) * (m * n + l) = \cos\theta * \sin\phi = \vec{K} * \vec{j} \quad [9.6]$$

$$\cos^2(V/2) * (1 - l^2 - m^2 + n^2) = \cos\theta * \cos\phi = \vec{K} * \vec{k}$$

By letting:

$$x_0 = \cos(V/2),\ x_1 = l * x_0,\ x_2 = m * x_0,\ x_3 = n * x_0 \quad [9.7]$$

we have:

$$\vec{I} * \vec{i} = x_0^2 + x_1^2 - x_2^2 - x_3^2 = 1 - 2 * (x_2^2 + x_3^2)$$

$$\vec{I} * \vec{j} = 2 * ((x_1 * x_2) - (x_0 * x_3))$$

$$\vec{I} * \vec{k} = 2 * ((x_1 * x_3) + (x_0 * x_2))$$

$$\vec{J} * \vec{i} = 2 * ((x_1 * x_2) + (x_0 * x_3))$$

$$\vec{J} * \vec{j} = 1 - 2 * (x_1^2 + x_3^2) \quad [9.8]$$

$$\vec{J} * \vec{k} = 2 * ((x_2 * x_3) - (x_0 * x_1))$$

$$\vec{K} * \vec{i} = 2 * ((x_1 * x_3) - (x_0 * x_2))$$

$$\vec{K} * \vec{j} = 2 * ((x_2 * x_3) + (x_0 * x_1))$$

$$\vec{K} * \vec{k} = 1 - 2 * (x_1^2 + x_2^2)$$

Having knowledge of every moment of the parameters (x_0, x_1, x_2, x_3) allows the Euler angles (Ψ, θ, ϕ) to be calculated using equations [9.4]–[9.6].

In order to determine the correspondence between the parameters (x_0, x_1, x_2, x_3) and the Euler angles (Ψ, θ, ϕ), it is interesting to go through a reference trihedron from the previous one (O, X, Y, Z) with a rotation of + 90° around the axis $\overrightarrow{OY}$. Let (O, X', Y', Z') be this new reference trihedron.

We have:

$$\vec{I'} = -\vec{K}, \quad \vec{J'} = \vec{J}, \quad \vec{K'} = \vec{I}$$

The unit vector of the instantaneous axis of rotation is written:

$$\vec{u} = \alpha * \vec{I} + \beta * \vec{J} + \gamma * \vec{K}$$

$$\vec{u} = -\gamma * \vec{I'} + \beta * \vec{J'} + \alpha * \vec{K'}$$

We obtain:

$$\vec{I'} * \vec{i} = 2 * ((x_0 * x_2) - (x_1 * x_3))$$

$$\vec{I'} * \vec{j} = -2 * ((x_0 * x_1) - (x_2 * x_3))$$

$$\vec{I'} * \vec{k} = 2 * ((x_0 * x_2) - (x_1 * x_3))$$

$$\vec{J'} * \vec{i} = 2 * ((x_1 * x_2) + (x_0 * x_3))$$

$$\vec{J'} * \vec{j} = 1 - 2 * (x_1^2 + x_3^2) \quad [9.9]$$

$$\vec{J'} * \vec{k} = 2 * ((x_2 * x_3) - (x_0 * x_1))$$

$$\vec{K'} * \vec{i} = 1 - 2 * (x_2^2 + x_3^2)$$

$$\vec{K'} * \vec{j} = 2 * ((x_1 * x_2) - (x_0 * x_3))$$

$$\vec{K'} * \vec{k} = 2 * ((x_1 * x_3) + (x_0 * x_2))$$

If (Ψ', θ', ϕ') designate the Euler angles in this new trihedron (O, X', Y', Z'), then we have the following identity:

$$1 - 2 * (x_2^2 + x_3^2) = -\sin(\theta') = \cos((\pi/2) + \theta')$$

Using the duplication formulas in trigonometry:

$$\cos(2a) = \cos^2(a) - \sin^2(a) = 2 * \cos^2(a) - 1 = 1 - 2\sin^2(a)$$

we obtain:

$$1 - 2 * (x_2^2 + x_3^2) = \cos\big((\pi/2) + \theta'\big) = 1 - 2 * \sin^2(((\pi/2) + \theta')/2)$$

Thus:

$$(x_2^2 + x_3^2) = \sin^2(((\pi/2) + \theta')/2)$$

by letting:

$$x_2 = a * \sin(((\pi/2) + \theta')/2) \text{ with } a^2 + b^2 = 1 \quad [9.10]$$

$$x_3 = b * \sin(((\pi/2) + \theta')/2)$$

From equation [9.8], we obtain:

$$x_0^2 + x_1^2 - x_2^2 - x_3^2 = 1 - 2 * (x_2^2 + x_3^2)$$

and thus:

$$x_0^2 + x_1^2 = 1 - 2 * (x_2^2 + x_3^2) + x_2^2 + x_3^2 = 1 - (x_2^2 + x_3^2)$$

$$x_0^2 + x_1^2 = 1 - \sin^2(((\pi/2) + \theta')/2) = \cos^2(((\pi/2) + \theta')/2)$$

by letting:

$$x_0 = c * \cos(((\pi/2) + \theta')/2)$$

with:

$$c^2 + d^2 = 1 \quad [9.11]$$

$$x_1 = d * \cos(((\pi/2) + \theta')/2)$$

By using these new expressions of (x_0, x_1, x_2, x_3) in equation [9.9] and by identifying them with the expressions of the directing cosines as a function of the angles (Ψ', θ', ϕ'), we obtain:

$$a * c - b * d = \cos \Psi' \Leftrightarrow a * c = \cos((\Psi' + \phi')/2) * \cos((\Psi' - \phi')/2)$$

$$a * d + b * d = \sin \Psi' \Leftrightarrow a * d = \cos((\Psi' + \phi')/2) * \cos((\Psi' - \phi')/2)$$

$$a * d - b * c = \sin \phi' \Leftrightarrow b * c = \cos((\Psi' + \phi')/2) * \sin((\Psi' - \phi')/2)$$

$$a * c + b * d = \cos \phi' \Leftrightarrow b * d = \sin((\Psi' + \phi')/2) * \sin((\Psi' - \phi')/2)$$

This results in:

$$a = \cos((\Psi' - \phi')/2)$$

$$b = \sin((\Psi' - \phi')/2)$$

$$c = \cos((\Psi' + \phi')/2)$$

$$d = \sin((\Psi' + \phi')/2)$$

From equations [9.10] and [9.11], we can deduce that:

$$x_0 = \cos(((\pi/2) + \theta')/2) * \cos((\Psi' + \phi')/2)$$

$$x_1 = \cos(((\pi/2) + \theta')/2) * \sin((\Psi' + \phi')/2)$$

$$x_2 = \sin(((\pi/2) + \theta')/2) * \cos((\Psi' - \phi')/2) \quad [9.12]$$

$$x_3 = \sin(((\pi/2) + \theta')/2) * \sin((\Psi' - \phi')/2)$$

Equation [9.12] is useful to initialize the parameters (x_0, x_1, x_2, x_3).

In Figure 9.2, α, β and γ represent the directing cosines of the instantaneous axis of rotation with respect to the trihedron (O, X, Y, Z) deduced from the trihedron (O, X', Y', Z'), with respect to which the angles are related (Ψ', θ', ϕ') by a rotation of (−π/2) around $\overrightarrow{OY'}$.

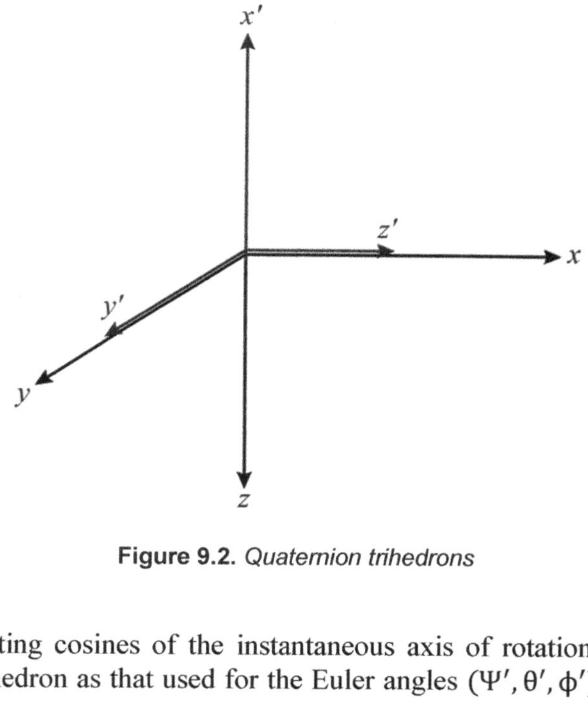

Figure 9.2. *Quaternion trihedrons*

The directing cosines of the instantaneous axis of rotation in the same reference trihedron as that used for the Euler angles (Ψ', θ', ϕ') are (−γ, +β, +α).

To obtain the differential equations which define the evolution over time of the terms (x0, x1, x2, x3) of the quaternion allowing the calculation of the change of axes, we can proceed as for the Euler method:

$$\frac{d\vec{I}}{dt} = \frac{d\vec{J}}{dt} = \frac{d\vec{K}}{dt} = 0$$

From equation [9.8], we can write:

$$\vec{I} = [1-2*(x_2^2+x_3^2)]*\vec{\imath}+2*\left((x_1*x_2)-(x_0*x_3)\right)*\vec{\jmath}+2*((x_1*x_3)+(x_0*x_2))*\vec{k}$$

$$\vec{J} = 2*((x_1*x_2)+(x_0*x_3))*\vec{\imath}+(1-2*(x_1^2+x_3^2))*\vec{\jmath}+2*((x_2*x_3)-(x_0*x_1))*\vec{k}$$

$$\vec{K} = 2*\left((x_1*x_3)-(x_0*x_2)\right)*\vec{\imath}+2*\left((x_2*x_3)+(x_0*x_1)\right)*\vec{\jmath}+\left[1-2*(x_1^2+x_2^2)\right]*\vec{k} \quad [9.13]$$

$$\frac{d\vec{\imath}}{dt} = \Omega \wedge \vec{\imath} = r*\vec{\jmath}-q*\vec{k}$$

$$\frac{d\vec{\jmath}}{dt} = \Omega \wedge \vec{\jmath} = (-r)*\vec{\imath}+p*\vec{k}$$

$$\frac{d\vec{k}}{dt} = \Omega \wedge \vec{k} = q*\vec{\imath}-p*\vec{\jmath} \quad [9.14]$$

By differentiating equation [9.13] and using the derivative equation $[(u*v)' = (u'*v)+(u*v')]$, we obtain:

$$\frac{d\vec{I}}{dt} = \frac{d[1-2*(x_2^2+x_3^2)]}{dt}*\vec{\imath}+[1-2*(x_2^2+x_3^2)]*\frac{d\vec{\imath}}{dt}+\frac{d(2*((x_1*x_2)-(x_0*x_3)))}{dt}*\vec{\jmath}+2*\left((x_1*x_2)-(x_0*x_3)\right)*\frac{d\vec{\jmath}}{dt}+\frac{d(2*((x_1*x_3)+(x_0*x_2)))}{dt}*\vec{k}+2*((x_1*x_3)+(x_0*x_2))*\frac{d\vec{k}}{dt}$$

$$\frac{d\vec{J}}{dt} = \frac{d(2*((x_1*x_2)+(x_0*x_3)))}{dt}*\vec{i} + (2*((x_1*x_2)+(x_0*x_3)))*\frac{d\vec{i}}{dt} + \frac{d\left((1-2*(x_1^2+x_3^2))\right)}{dt}*\vec{j} + \left((1-2*(x_1^2+x_3^2))\right)*\frac{d\vec{j}}{dt} + \frac{d(2*((x_2*x_3)-(x_0*x_1)))}{dt}*\vec{k} + (2*((x_2*x_3)-(x_0*x_1)))*\frac{d\vec{k}}{dt}$$

$$\frac{d\vec{K}}{dt} = \frac{d(2*((x_1*x_3)-(x_0*x_2)))}{dt}*\vec{i} + (2*((x_1*x_3)-(x_0*x_2)))*\frac{d\vec{i}}{dt} + \frac{d(2*((x_2*x_3)+(x_0*x_1)))}{dt}*\vec{j} + (2*((x_2*x_3)+(x_0*x_1)))*\frac{d\vec{j}}{dt} + \frac{d(1-2*(x_1^2+x_2^2))}{dt}*\vec{k} + (1-2*(x_1^2+x_2^2))*\frac{d\vec{k}}{dt}$$

According to equation [9.14], we obtain:

$$0 = \frac{d\vec{I}}{dt} = (-4*(x_2*\dot{x}_2+x_3*\dot{x}_3)*\vec{i} + [1-2*(x_2^2+x_3^2)]*(r*\vec{j} - q*\vec{k}) + \left(2*\left((\dot{x}_1*x_2)+(x_1*\dot{x}_2)-(\dot{x}_0*x_3)-(x_0*\dot{x}_3)\right)\right)*\vec{j} + \left(2*\left((x_1*x_2)-(x_0*x_3)\right)\right)*\left((-r)*\vec{i}+p*\vec{k}\right) + (2*((\dot{x}_1*x_3)+(x_1*\dot{x}_3)+(\dot{x}_0*x_2)+(x_0*\dot{x}_2)))*\vec{k} + 2*((x_1*x_3)+(x_0*x_2))*(q*\vec{i}-p*\vec{j})$$

The three components following $\vec{i}, \vec{j}$ and $\vec{k}$ must be zero.

Focusing on the axis $(O,\vec{i})$, let:

$$\left(\left(-4 * (x_2 * \dot{x}_2 + x_3 * \dot{x}_3) + 2 * (-r) * \left((x_1 * x_2) - (x_0 * x_3)\right) + (2 * ((x_1 * x_3) + (x_0 * x_2)) * q\right) * \vec{i} = 0$$

and let:

$$(x_2 * \dot{x}_2 + x_3 * \dot{x}_3) = \left(\frac{q}{2}\right) * \left((x_1 * x_3) + (x_0 * x_2)\right) - \left(\frac{r}{2}\right) * \left((x_1 * x_2) - (x_0 * x_3)\right) \quad [9.15]$$

$$\begin{aligned} 0 = \frac{d\vec{j}}{dt} = {} & (2 * ((\dot{x}_1 * x_2) + (x_1 * \dot{x}_2) + (\dot{x}_0 * x_3) + (x_0 * \dot{x}_3))) * \vec{i} \\ & + \left(2 * \left((x_1 * x_2) + (x_0 * x_3)\right)\right) * (r * \vec{j} - q * \vec{k}) \\ & - \left(4 * (x_1 * \dot{x}_1) + (x_3 * \dot{x}_3)\right) * \vec{j} \\ & + \left(\left(1 - 2 * (x_1^2 + x_3^2)\right)\right) * ((-r) * \vec{i} + p * \vec{k}) + (2 \\ & * ((\dot{x}_2 * x_3) + (x_2 * \dot{x}_3) - (\dot{x}_0 * x_1) - (x_0 * \dot{x}_1))) * \vec{k} \\ & + (2 * ((x_2 * x_3) - (x_0 * x_1))) * (q * \vec{i} - p * \vec{j}) \end{aligned}$$

Focusing on the axis $(O,\vec{j})$, let:

$$\left[\left(2 * \left((x_1 * x_2) + (x_0 * x_3)\right)\right) * r - \left(4 * (x_1 * \dot{x}_1) + (x_3 * \dot{x}_3)\right) - \left(2 * \left((x_2 * x_3) - (x_0 * x_1)\right)\right) * p\right] * \vec{j} = 0$$

and let:

$$(x_1 * \dot{x}_1) + (x_3 * \dot{x}_3) = \left((x_1 * x_2) + (x_0 * x_3)\right) * \left(\frac{r}{2}\right) - \left((x_2 * x_3) - (x_0 * x_1)\right) * \left(\frac{p}{2}\right) \quad [9.16]$$

$$0 = \frac{d\vec{K}}{dt} = (2 * ((\dot{x}_1 * x_3) + (x_1 * \dot{x}_3) - (\dot{x}_0 * x_2) - (x_0 * \dot{x}_2))) * \vec{i}$$
$$+ \left(2 * ((x_1 * x_3) - (x_0 * x_2))\right) * (r * \vec{j} - q * \vec{k})$$
$$+ \left(2 * ((\dot{x}_0 * x_1) + (x_0 * \dot{x}_1) + (\dot{x}_2 * x_3) + (x_2 * \dot{x}))\right) * \vec{j}$$
$$+ \left(2 * ((x_0 * x_1) + (x_2 * x_3))\right) * \left((-r) * \vec{i} + p * \vec{k}\right)$$
$$- 4 * (x_1 * \dot{x}_1 + x_2 * \dot{x}_2) * \vec{k} + (1 - 2 * (x_1^2 + x_2^2)) * (q * \vec{i} - p * \vec{j})$$

Focusing on the axis $(O,\vec{k})$, let:

$$-\left(2 * ((x_1 * x_3) - (x_0 * x_2))\right) * q + \left(2 * ((x_0 * x_1) + (x_2 * x_3))\right) * p - 4 * (x_1 * \dot{x}_1 + x_2 * \dot{x}_2) = 0$$

and let:

$$(x_1 * \dot{x}_1 + x_2 * \dot{x}_2) = ((x_0 * x_2) - (x_1 * x_3)) * \left(\frac{q}{2}\right) + ((x_0 * x_1) + (x_2 * x_3)) * \left(\frac{p}{2}\right) \quad [9.17]$$

By deriving the equation $x_0^2 + x_1^2 - x_2^2 - x_3^2 = 1$, we obtain:

$$x_0 * \dot{x}_0 + x_1 * \dot{x}_1 + x_2 * \dot{x}_2 + x_3 * \dot{x}_3 = 0 \quad [9.18]$$

The differential system is written:

$$x_1 * \dot{x}_1 + x_2 * \dot{x}_2 = ((x_0 * x_2) - (x_1 * x_3)) * \left(\frac{q}{2}\right) + ((x_0 * x_1) + (x_2 * x_3)) * \left(\frac{p}{2}\right)$$

$$x_1 * \dot{x}_1 + x_3 * \dot{x}_3 = ((x_1 * x_2) + (x_0 * x_3)) * \left(\frac{r}{2}\right) - ((x_2 * x_3) - (x_0 * x_1)) * \left(\frac{p}{2}\right)$$

$$
\begin{aligned}
(x_2 * \dot{x}_2 + x_3 * \dot{x}_3) \\
&= \left(\frac{q}{2}\right) * \left((x_1 * x_3) + (x_0 * x_2)\right) \\
&- \left(\frac{r}{2}\right) \qquad\qquad * \left((x_1 * x_2) - (x_0 * x_3)\right)
\end{aligned}
$$

$$
x_0 * \dot{x}_0 + x_1 * \dot{x}_1 + x_2 * \dot{x}_2 + x_3 * \dot{x}_3 = 0 \qquad [9.19]
$$

By summing the first three parts of equation [9.19] and comparing them to the fourth part, we obtain:

$$
\begin{aligned}
x_1 * \dot{x}_1 + x_2 * \dot{x}_2 + x_1 * \dot{x}_1 + x_3 * \dot{x}_3 + x_2 * \dot{x}_2 + x_3 * \dot{x}_3 \\
&= \left((x_0 * x_2) - (x_1 * x_3)\right) * \left(\frac{q}{2}\right) + (x_0 * x_1) * \left(\frac{p}{2}\right) \\
&+ (x_2 * x_3) * \left(\frac{p}{2}\right) + (x_1 * x_2) * \left(\frac{r}{2}\right) + (x_0 * x_3) * \left(\frac{r}{2}\right) \\
&- (x_2 * x_3) * \left(\frac{p}{2}\right) + (x_0 * x_1) * \left(\frac{p}{2}\right) + \left(\frac{q}{2}\right) * (x_1 * x_3) \\
&+ \left(\frac{q}{2}\right) * (x_0 * x_2) - \left(\frac{r}{2}\right) * (x_1 * x_2) + \left(\frac{r}{2}\right) * (x_0 * x_3)
\end{aligned}
$$

$$
\begin{aligned}
2 * (x_1 * \dot{x}_1) + 2 * (x_2 * \dot{x}_2) + 2 * (x_3 * \dot{x}_3) \\
&= (x_0 * x_2) * \left(\frac{q}{2}\right) - (x_1 * x_3) * \left(\frac{q}{2}\right) + (x_0 * x_1) \\
&* \left(\frac{p}{2}\right) + (x_2 * x_3) * \left(\frac{p}{2}\right) + (x_1 * x_2) * \left(\frac{r}{2}\right) + (x_0 * x_3) \\
&* \left(\frac{r}{2}\right) - (x_2 * x_3) * \left(\frac{p}{2}\right) + (x_0 * x_1) * \left(\frac{p}{2}\right) + \left(\frac{q}{2}\right) \\
&* (x_1 * x_3) + \left(\frac{q}{2}\right) * (x_0 * x_2) - \left(\frac{r}{2}\right) * (x_1 * x_2) + \left(\frac{r}{2}\right) \\
&* (x_0 * x_3)
\end{aligned}
$$

We then obtain:

$$
\begin{aligned}
2 * (x_1 * \dot{x}_1) + 2 * (x_2 * \dot{x}_2) + 2 * (x_3 * \dot{x}_3) \\
&= (x_0 * x_2) * \left(\frac{q}{2}\right) + (x_0 * x_1) * \left(\frac{p}{2}\right) + (x_0 * x_3) \\
&* \left(\frac{r}{2}\right) + (x_0 * x_1) * \left(\frac{p}{2}\right) + \left(\frac{q}{2}\right) * (x_0 * x_2) + \left(\frac{r}{2}\right) \\
&* (x_0 * x_3) \\
&= 2 * (x_0 * x_2) * \left(\frac{q}{2}\right) + 2 * (x_0 * x_1) * \left(\frac{p}{2}\right) + 2 \\
&* (x_0 * x_3) * \left(\frac{r}{2}\right)
\end{aligned}
$$

and:

$$2 * (x_1 * \dot{x}_1) + 2 * (x_2 * \dot{x}_2) + 2 * (x_3 * \dot{x}_3) = 2 * (x_0 * x_2) * \left(\frac{q}{2}\right) + 2 * (x_0 * x_1) * \left(\frac{p}{2}\right) + 2 * (x_0 * x_3) * \left(\frac{r}{2}\right)$$

Thus:

$$(x_1 * \dot{x}_1) + (x_2 * \dot{x}_2) + (x_3 * \dot{x}_3) = \frac{1}{2} * [(x_0 * x_2) * q + 2 * (x_0 * x_1) * p + 2 * (x_0 * x_3) * r] = -(x_0 * \dot{x}_0)$$

$$\dot{x}_0 = -\frac{1}{2} * [x_1 * p + x_2 * q + x_3 * r] \quad [9.20]$$

Now that we know x_0, we need to find $\dot{x}_1, \dot{x}_2$ and $\dot{x}_3$.

We add the first and the second parts of equation [9.19] and subtract the third part:

$$(x_1 * \dot{x}_1 + x_2 * \dot{x}_2) + (x_1 * \dot{x}_1 + x_3 * \dot{x}_3) - (x_2 * \dot{x}_2 + x_3 * \dot{x}_3) = \left(\left((x_0 * x_2) - (x_1 * x_3)\right) * \left(\frac{q}{2}\right) + \left((x_0 * x_1) + (x_2 * x_3)\right) * \left(\frac{p}{2}\right)\right) + \left(\left((x_1 * x_2) + (x_0 * x_3)\right) * \left(\frac{r}{2}\right) - \left((x_2 * x_3) - (x_0 * x_1)\right) * \left(\frac{p}{2}\right)\right) - \left(\left(\frac{q}{2}\right) * \left((x_1 * x_3) + (x_0 * x_2)\right) - \left(\frac{r}{2}\right) * \left((x_1 * x_2) - (x_0 * x_3)\right)\right)$$

$$\begin{aligned}(x_1 * \dot{x}_1 + x_2 * \dot{x}_2) + (x_1 * \dot{x}_1 + x_3 * \dot{x}_3) - (x_2 * \dot{x}_2 + x_3 * \dot{x}_3) \\ = (x_0 * x_2) * \left(\frac{q}{2}\right) - (x_1 * x_3) * \left(\frac{q}{2}\right) + (x_0 * x_1) \\ * \left(\frac{p}{2}\right) + (x_2 * x_3) * \left(\frac{p}{2}\right) + (x_1 * x_2) * \left(\frac{r}{2}\right) + (x_0 * x_3) \\ * \left(\frac{r}{2}\right) - (x_2 * x_3) * \left(\frac{p}{2}\right) + (x_0 * x_1) * \left(\frac{p}{2}\right) - (x_1 * x_3) \\ * \left(\frac{q}{2}\right) - (x_0 * x_2) * \left(\frac{q}{2}\right) + (x_1 * x_2) * \left(\frac{r}{2}\right) \\ - (x_0 * x_3)\left(\frac{r}{2}\right)\end{aligned}$$

$$\begin{aligned}(x_1 * \dot{x}_1 + x_2 * \dot{x}_2) + (x_1 * \dot{x}_1 + x_3 * \dot{x}_3) - (x_2 * \dot{x}_2 + x_3 * \dot{x}_3) \\ = -(x_1 * x_3) * \left(\frac{q}{2}\right) + (x_0 * x_1) * \left(\frac{p}{2}\right) + (x_1 * x_2) \\ * \left(\frac{r}{2}\right) + (x_0 * x_1) * \left(\frac{p}{2}\right) - (x_1 * x_3) * \left(\frac{q}{2}\right) + (x_1 * x_2) \\ * \left(\frac{r}{2}\right)\end{aligned}$$

$$2 * (x_1 * \dot{x}_1) = p * (x_0 * x_1) + r * (x_1 * x_2) - q * (x_1 * x_3)$$

We then add the first and the third parts of equation [9.19] and subtract the second part:

$$\begin{aligned}(x_1 * \dot{x}_1 + x_2 * \dot{x}_2) - (x_1 * \dot{x}_1 + x_3 * \dot{x}_3) + (x_2 * \dot{x}_2 + x_3 * \dot{x}_3) \\ = \Bigg(\Big((x_0 * x_2) - (x_1 * x_3)\Big) * \left(\frac{q}{2}\right) \\ + \Big((x_0 * x_1) + (x_2 * x_3)\Big) * \left(\frac{p}{2}\right)\Bigg) \\ - \Bigg(\Big((x_1 * x_2) + (x_0 * x_3)\Big) * \left(\frac{r}{2}\right) \\ - \Big((x_2 * x_3) - (x_0 * x_1)\Big) * \left(\frac{p}{2}\right)\Bigg) \\ + \Bigg(\left(\frac{q}{2}\right) * \Big((x_1 * x_3) + (x_0 * x_2)\Big) - \left(\frac{r}{2}\right) \\ * \Big((x_1 * x_2) - (x_0 * x_3)\Big)\Bigg)\end{aligned}$$

$$
\begin{aligned}
&(x_1 * \dot{x}_1 + x_2 * \dot{x}_2) - (x_1 * \dot{x}_1 + x_3 * \dot{x}_3) + (x_2 * \dot{x}_2 + x_3 * \dot{x}_3) \\
&\quad = (x_0 * x_2) * \left(\frac{q}{2}\right) - (x_1 * x_3) * \left(\frac{q}{2}\right) + (x_0 * x_1) \\
&\quad * \left(\frac{p}{2}\right) + (x_2 * x_3) * \left(\frac{p}{2}\right) - (x_1 * x_2) * \left(\frac{r}{2}\right) - (x_0 * x_3) \\
&\quad * \left(\frac{r}{2}\right) + (x_2 * x_3) * \left(\frac{p}{2}\right) - (x_0 * x_1) * \left(\frac{p}{2}\right) + (x_1 * x_3) \\
&\quad * \left(\frac{q}{2}\right) + (x_0 * x_2) * \left(\frac{q}{2}\right) - (x_1 * x_2) * \left(\frac{r}{2}\right) + (x_0 * x_3) \\
&\quad * \left(\frac{r}{2}\right)
\end{aligned}
$$

$$
\begin{aligned}
&2 * (x_2 * \dot{x}_2) \\
&\quad = (x_0 * x_2) * \left(\frac{q}{2}\right) + (x_2 * x_3) * \left(\frac{p}{2}\right) - (x_1 * x_2) \\
&\quad * \left(\frac{r}{2}\right) + (x_2 * x_3) * \left(\frac{p}{2}\right) + (x_0 * x_2) * \left(\frac{q}{2}\right) - (x_1 * x_2) \\
&\quad * \left(\frac{r}{2}\right)
\end{aligned}
$$

$$
2 * (x_2 * \dot{x}_2) = q * (x_0 * x_2) - r * (x_1 * x_2) + p * (x_2 * x_3)
$$

We then add the second and third parts of equation [9.19] and subtract the first part:

$$
\begin{aligned}
&-(x_1 * \dot{x}_1 + x_2 * \dot{x}_2) + (x_1 * \dot{x}_1 + x_3 * \dot{x}_3) + (x_2 * \dot{x}_2 + x_3 * \dot{x}_3) \\
&\quad = -\Bigg(\Big((x_0 * x_2) - (x_1 * x_3)\Big) * \left(\frac{q}{2}\right) \\
&\quad + \Big((x_0 * x_1) + (x_2 * x_3)\Big) * \left(\frac{p}{2}\right)\Bigg) \\
&\quad + \Bigg(\Big((x_1 * x_2) + (x_0 * x_3)\Big) * \left(\frac{r}{2}\right) \\
&\quad - \Big((x_2 * x_3) - (x_0 * x_1)\Big) * \left(\frac{p}{2}\right)\Bigg) \\
&\quad + \Bigg(\left(\frac{q}{2}\right) * \Big((x_1 * x_3) + (x_0 * x_2)\Big) - \left(\frac{r}{2}\right) \\
&\quad * \Big((x_1 * x_2) - (x_0 * x_3)\Big)\Bigg)
\end{aligned}
$$

$$-(x_1 * \dot{x}_1 + x_2 * \dot{x}_2) + (x_1 * \dot{x}_1 + x_3 * \dot{x}_3) + (x_2 * \dot{x}_2 + x_3 * \dot{x}_3)$$
$$= -(x_0 * x_2) * \left(\frac{q}{2}\right) + (x_1 * x_3) * \left(\frac{q}{2}\right) - (x_0 * x_1) * \left(\frac{p}{2}\right) - (x_2 * x_3) * \left(\frac{p}{2}\right) + (x_1 * x_2) * \left(\frac{r}{2}\right) + (x_0 * x_3) * \left(\frac{r}{2}\right) - (x_2 * x_3) * \left(\frac{p}{2}\right) + (x_0 * x_1) * \left(\frac{p}{2}\right) + (x_1 * x_3) * \left(\frac{q}{2}\right) + (x_0 * x_2) * \left(\frac{q}{2}\right) - (x_1 * x_2) * \left(\frac{r}{2}\right) + (x_0 * x_3) * \left(\frac{r}{2}\right)$$

$$-(x_1 * \dot{x}_1 + x_2 * \dot{x}_2) + (x_1 * \dot{x}_1 + x_3 * \dot{x}_3) + (x_2 * \dot{x}_2 + x_3 * \dot{x}_3) + (x_1 * x_3) * \left(\frac{q}{2}\right) - (x_2 * x_3) * \left(\frac{p}{2}\right) + (x_0 * x_3) * \left(\frac{r}{2}\right) - (x_2 * x_3) * \left(\frac{p}{2}\right) + (x_1 * x_3) * \left(\frac{q}{2}\right) + (x_0 * x_3) * \left(\frac{r}{2}\right)$$

$$2 * (x_3 * \dot{x}_3) = r * (x_0 * x_3) + q * (x_1 * x_3) - p * (x_2 * x_3)$$

In summary:

$$2 * (x_1 * \dot{x}_1) = p * (x_0 * x_1) + r * (x_1 * x_2) - q * (x_1 * x_3)$$

$$2 * (x_2 * \dot{x}_2) = q * (x_0 * x_2) - r * (x_1 * x_2) + p * (x_2 * x_3)$$

$$2 * (x_3 * \dot{x}_3) = r * (x_0 * x_3) + q * (x_1 * x_3) - p * (x_2 * x_3)$$

This system of three equations is equivalent to:

$$\dot{x}_1 = \frac{1}{2} * (p * x_0 + r * x_2 - q * x_3)$$

$$\dot{x}_2 = \frac{1}{2} * (q * x_0 - r * x_1 + p * x_3)$$

$$\dot{x}_3 = \frac{1}{2} * (r * x_0 + q * x_1 - p * x_2) \qquad [9.21]$$

Equations [9.20] and [9.21] make it possible to use step-by-step numerical integration to obtain the new values of the four parameters (x_0, x_1, x_2, x_3) at each step of the calculation.

NOTE.– The four parameters (x_0, x_1, x_2, x_3) are not independent. In accordance with the first part of equation [9.8], we have $x_0^2 + x_1^2 + x_2^2 + x_3^2 = 1$.

The inaccuracies resulting from the separate integration of the four differential equations have to be corrected by possibly "renormalizing" the four parameters at each calculation step.

The four differential equations are of the first order, with non-constant coefficients, but they do not involve the trigonometric lines.

There is no longer a problem of angular determination when the three Euler angles (Ψ, θ, ϕ) approach or pass through 0° and/or 90°. The physical representation of the parameters (x_0, x_1, x_2, x_3) is highlighted, because:

$$x_0 = \cos\frac{V}{2}$$

$$x_1 = \alpha * \sin\frac{V}{2}$$

$$x_2 = \beta * \sin\frac{V}{2}$$

$$x_3 = \gamma * \sin\frac{V}{2}$$

α, β and γ represent the directing cosines of the instantaneous axis of rotation with respect to the trihedron (O, X, Y, Z) deduced from the trihedron (O, X', Y', Z'), and V is the instantaneous amplitude of the rotation (in the reference trihedron of Euler angles (Ψ, θ, ϕ). The directing cosines of the instantaneous axis of rotation are $(-\gamma, +\beta, +\alpha)$).

Finally, with (Q) denoting the unitary quaternion defined by (x_0, x_1, x_2, x_3) and (ω) the quaternion representing the instantaneous vector $(\vec{\Omega})$ of rotation, we can write:

$$(\dot{Q}) = \frac{dQ}{dt} = \frac{1}{2} * (Q) * (\omega)$$

with:

$$(Q) = x_0 + x_1 * \vec{i} + x_2 * \vec{j} + x_3 * \vec{k} et x_0^2 + x_1^2 + x_2^2 + x_3^2 = 1$$

$$(\omega) = \left(0, \vec{\Omega}\right) = p * \vec{i} + q * \vec{j} + r * \vec{k}$$

$$\left(\dot{Q}Q\right) = \dot{x}_0 + \dot{x}_1 * \vec{i} + \dot{x}_2 * \vec{j} + \dot{x}_3 * \vec{k}$$

See Laisant (1877) and De Casteljau (1997).

Glossary

Aerodynamic drag: the component of the aerodynamic force generated by the reaction of the surrounding gas of a moving body, in the direction of the relative wind. A distinction is made between frictional drag, wave drag (shock wave at the front) and base drag (depression at the rear).

Aerodynamic focus: the point of application, on the axis of an aerospace machine, of the resultant aerodynamic forces of lift and drag.

Aircraft: a vehicle whose lift in the atmosphere is provided by the reaction of the air.

Aircraft load factor (ng): the aircraft acceleration per g (acceleration of gravity – (9.81 m/s^2).

Angle of incidence or angle of attack: the algebraic value of the angle formed by the projection of the axis OX_{va} of the aerodynamic trihedron on the plane of symmetry X_eOZ_e of the gear trihedron and the reference axis OX_e of this last trihedron.

Boundary layer: a layer of a moving fluid in the vicinity of a wall in which, due to the viscosity of the fluid, the free flow equations are not directly applicable.

Center of gravity: the point of application of the resultant of the external forces exerted on all parts of a body by a gravitational field.

Center of inertia: the point of application of the resultant of the inertial forces of a body subjected to an acceleration field.

Center of mass: the barycenter of the elementary masses of a body.

Center of thrust: the point of application of the resultant of aerodynamic forces.

Control chain: a set of equipment that ensures the development of the piloting orders of a machine and their transformation into mechanical action.

Correlation: the interdependence between quantitative or qualitative observations, highlighted by statistical analysis.

Detachment: the characteristic of a fluid flow that departs from the wall it runs along.

Drag: a component, following the direction of speed, of the forces that oppose the advancement of the aircraft.

Euler angles: Euler angles define the position of the aircraft trihedron M with respect to the local geographic trihedron L in which the vector g has as components (0, 0, g).

Geoid: a representation of the Earth's surface that is more precise than the spherical or ellipsoidal approximation. It corresponds to an equipotential (in the field of terrestrial gravity) and is defined in such a way as to stick as closely as possible to the "real area". On Earth, any point undergoes an acceleration of gravity g. This acceleration derives from a gravitational potential W, such that $\vec{g} = \overrightarrow{grad}(W)$. The surfaces where the potential of gravity W is constant are equipotentials of gravity. A geoid is an equipotential surface of gravity close to the mean sea level.

Gravitation: a physical phenomenon whereby bodies attract each other in proportion to their masses and inversely to the square of their distance.

Gravity: the gravitational force exerted by an astral body on another body.

Hinge moment: the moment resulting from aerodynamic forces exerted on a steerable control surface, with respect to the axis of rotation of the latter.

Inertial unit: a precision mechanical and electronic assembly comprising sensitive components (accelerometers, gyrometers) as well as calculation means, capable of supplying the position coordinates and components of speed and acceleration of the machine at any time.

Leading edge: the anterior part of a wing profile where the airflow divides into the lower surface flow and the upper surface flow.

Lift: a component, perpendicular to the relative wind, of the aerodynamic force generated by the reaction of the surrounding gas of a moving body.

Navigation: the art of driving an aeronautical vehicle to a given destination by determining its position, calculating the optimal trajectory and guiding it.

Piloting: the act of controlling the orientation and the speed of an aeronautical vehicle.

Pitch: the movement of a body around the pitch axis, perpendicular to the roll axis (for an airplane, the pitch axis is by convention an axis perpendicular to its plane of symmetry).

Relative wind: the speed of a gaseous fluid relative to a body immersed in it.

Remote control: the elaboration and remote transmission of a signal carrying an order.

Roll: the movement of a body around the so-called roll axis (for an airplane, a drone or a rocket, the roll axis is by convention the longitudinal axis).

Rotation: the circular movement of a body around an axis passing through the body's center of mass.

Steering: the terminal part of a control chain, which acts on the flow of the ejected fluid or the surrounding fluid.

Structure: a set of rigid elements that determine the shape of an aircraft and its ability to withstand mechanical and thermal stresses.

System: a functional whole the parts of which are interconnected and exchange matter, energy or information.

Telemetry: the elaboration and remote transmission of a signal carrying a measurement result.

Thrust axis: the axis carrying the resultant of the propulsive forces.

Tracking: the remote, instantaneous and continuous determination of one or more characteristic variables of an aircraft generally linked to its movement.

Trailing edge: the rear part of a wing profile where the intrados and extrados flows meet.

Trajectography: an operation that consists of tracing or subsequently reconstituting the trajectory of an aircraft.

Trajectory: a curve described by the center of mass of a moving body.

Wind tunnel: a test installation in which the flow of a gas makes it possible to study the aerodynamic or ballistic behavior of a body placed in this flow.

Yaw: the movement of a body around the yaw axis, perpendicular to the pitch and roll axes.

List of Abbreviations

(x_c, y_c, z_c)	Aircraft coordinates in the trajectography reference trihedron
z_{cgeoid}	Aircraft altitude above the geoid (trajectography)
(x_0, y_0, z_0) (O, X_0, Y_0, Z_0)	Aircraft coordinates in the calculation reference trihedron
z_{0geoid}	Aircraft altitude above the geoid
Ψ_{0c} and ϕ_{0c}	Angles allowing the passage from the trihedron (O, X_C, Y_c, Z_c) to the trihedron (O, X_0, Y_0, Z_0)
Ψ	Aircraft heading
Ψ_0	Reference heading
Ψ_c	Lateral attitude of the aircraft in relation to the reference heading
χ_a	Aerodynamic route of the aircraft
χ_{ac}	Aerodynamic course of the aircraft in relation to the reference heading

γ_a	Aerodynamic slope of the aircraft in relation to the calculation reference trihedron
γ_{avl}	Aerodynamic slope of the aircraft in relation to the local horizontal plane (OX_{vl}, OY_{vl})
α and β	Aircraft angle of attack and sideslip
α_t and i_t	Total impact of the aircraft
θ	Aircraft vertical attitude
θ_{vl}	Vertical attitude of the aircraft in relation to the local horizontal plane
ϕ	Aircraft roll
ϕ_a	Aerodynamic roll
R	Radius of curvature of the Earth's geoid
L	Reference length
S	Reference surface
M	Aircraft mass
V_v and C_v	Wind speed and direction
V	Aircraft ground speed
V_a	Aerodynamic speed of the aircraft
P and T	Pressure and temperature
M	Mach number
nx_a, ny_{af}, nz_r	Aircraft load factors in the trihedron (O, X_a, Y_{af}, Z_r)
nx_1, ny_1, nz_1	Aircraft load factors in the aircraft trihedron

C_y and C_z	Lateral and normal lift coefficients in the aircraft trihedron
C_{yr}	Coefficients of the normal damping forces due to the yaw rate
C_{zq}	Coefficients of normal damping forces due to pitching speed
η, ζ, ξ	Depth, direction and roll orders
$\sigma_1, \sigma_2, \sigma_3, \sigma_4$	Control surface deflections
p_e, q_e, $r_e e$	Components of the rotation vector in the aircraft trihedron
C_x	Coefficient of drag
C_0	Minimum drag coefficient
ΔC_{x1}	Coefficient of induced drag
ΔC_{x2}	Control surface drag coefficient
F_p	Propulsion force

References

Annequin, R. and Boutigny, J. (1974). *Cours de sciences physiques mécanique 2*, 2nd edition. Librairie Vuibert, Paris.

Bruhat, G. (1967). *Cours de physique générale, mécanique*, 6th edition. Masson & Cie, Paris.

Coriolis, G.G. (1835). Mémoire sur les équations du mouvement relatif des systèmes de corps. *Journal de l'école polytechnique*, 15(1835), 142–154.

Cousteix, J. and Gouverneur, C. (2016). *Aérodynamique physique et concepts de base*. Cépaduès Éditions, Toulouse.

De Casteljau, P. (1997). *Les Quaternions*. Hermes, Paris.

Délery, J. (2017). *Aérodynamique expérimentale. Souffleries et méthodes de mesure*. Cépaduès Éditions, Toulouse.

Direction générale de l'aviation civile (2012). *Le décrochage revenir aux incidences de vol*. Direction de la sécurité de l'aviation civile, Paris.

Giovannini, A. and Airiau, C. (2016). *Aérodynamique fondamentale*. Cépaduès Éditions, Toulouse.

Harcaut, J.P., Larcher, E., Bonnet, J.P., Dupont, S. (1998). Considérations sur l'aérodynamique pour le pilotage et le guidage des engins tactiques. *Aérospatiale missiles*, 5(30).

Kermode, A.C. (2012). *Mechanics of Flight*. Pearson Eduction, London.

Klopfstein, G. (2008). *Comprendre l'avion : Tome 2, Mécanique du vol*. Cépaduès Éditions, Toulouse.

Laisant, M. (1877). Applications mécaniques du calcul des quaternions. PhD thesis, Faculté des Sciences de Paris, Paris.

Landau, L. and Lifchitz, E. (1998). *Physique théorique – mécanique des fluides*, 3rd edition. Eyrolles, Paris.

Luneau, J. (1974). *Tables de détente ou compression isentropique et tables de choc γ = 1,400*. Cépaduès Éditions, Toulouse.

Index

Other titles from

iSTE

in

Systems and Industrial Engineering – Robotics

2022

HAJJI Rafika, JARAR OULIDI Hassane
Building Information Modeling for a Smart and Sustainable Urban Space

2020

BRON Jean-Yves
System Requirements Engineering

KRYSINSKI TOMASZ, MALBURET FRANÇOIS
Energy and Motorization in the Automotive and Aeronautics Industries

PRINTZ Jacques
System Architecture and Complexity: Contribution of Systems of Systems to Systems Thinking

2019

ANDRÉ Jean-Claude
Industry 4.0: Paradoxes and Conflicts

BENSALAH Mounir, ELOUADI Abdelmajid, MHARZI Hassan
Railway Information Modeling RIM: The Track to Rail Modernization

BLUA Philippe, YALAOU Farouk, AMODEO Lionel, DE BLOCK Michaël, LAPLANCHE David
Hospital Logistics and e-Management: Digital Transition and Revolution

BRIFFAUT Jean-Pierre
From Complexity in the Natural Sciences to Complexity in Operations Management Systems
(Systems of Systems Complexity Set – Volume 1)

BUDINGER Marc, HAZYUK Ion, COÏC Clément
Multi-Physics Modeling of Technological Systems

FLAUS Jean-Marie
Cybersecurity of Industrial Systems

JAULIN Luc
Mobile Robotics – Second Edition Revised and Updated

KUMAR Kaushik, DAVIM Paulo J.
Optimization for Engineering Problems

TRIGEASSOU Jean-Claude, MAAMRI Nezha
Analysis, Modeling and Stability of Fractional Order Differential Systems 1: The Infinite State Approach
Analysis, Modeling and Stability of Fractional Order Differential Systems 2: The Infinite State Approach

VANDERHAEGEN Frédéric, MAAOUI Choubeila, SALLAK Mohamed, BERDJAG Denis
Automation Challenges of Socio-technical Systems

2018

BERRAH Lamia, CLIVILLÉ Vincent, FOULLOY Laurent
Industrial Objectives and Industrial Performance: Concepts and Fuzzy Handling

GONZALEZ-FELIU Jesus
Sustainable Urban Logistics: Planning and Evaluation

GROUS Ammar
Applied Mechanical Design

LEROY Alain
Production Availability and Reliability: Use in the Oil and Gas Industry

MARÉ Jean-Charles
Aerospace Actuators 3: European Commercial Aircraft and Tiltrotor Aircraft

MAXA Jean-Aimé, BEN MAHMOUD Mohamed Slim, LARRIEU Nicolas
Model-driven Development for Embedded Software: Application to Communications for Drone Swarm

MBIHI Jean
Analog Automation and Digital Feedback Control Techniques
Advanced Techniques and Technology of Computer-Aided Feedback Control

MORANA Joëlle
Logistics

SIMON Christophe, WEBER Philippe, SALLAK Mohamed
Data Uncertainty and Important Measures
(Systems Dependability Assessment Set – Volume 3)

TANIGUCHI Eiichi, THOMPSON Russell G.
City Logistics 1: New Opportunities and Challenges
City Logistics 2: Modeling and Planning Initiatives
City Logistics 3: Towards Sustainable and Liveable Cities

ZELM Martin, JAEKEL Frank-Walter, DOUMEINGTS Guy, WOLLSCHLAEGER Martin
Enterprise Interoperability: Smart Services and Business Impact of Enterprise Interoperability

2017

ANDRÉ Jean-Claude
From Additive Manufacturing to 3D/4D Printing 1: From Concepts to Achievements

From Additive Manufacturing to 3D/4D Printing 2: Current Techniques, Improvements and their Limitations
From Additive Manufacturing to 3D/4D Printing 3: Breakthrough Innovations: Programmable Material, 4D Printing and Bio-printing

ARCHIMÈDE Bernard, VALLESPIR Bruno
Enterprise Interoperability: INTEROP-PGSO Vision

CAMMAN Christelle, FIORE Claude, LIVOLSI Laurent, QUERRO Pascal
Supply Chain Management and Business Performance: The VASC Model

FEYEL Philippe
Robust Control, Optimization with Metaheuristics

MARÉ Jean-Charles
Aerospace Actuators 2: Signal-by-Wire and Power-by-Wire

POPESCU Dumitru, AMIRA Gharbi, STEFANOIU Dan, BORNE Pierre
Process Control Design for Industrial Applications

RÉVEILLAC Jean-Michel
Modeling and Simulation of Logistics Flows 1: Theory and Fundamentals
Modeling and Simulation of Logistics Flows 2: Dashboards, Traffic Planning and Management
Modeling and Simulation of Logistics Flows 3: Discrete and Continuous Flows in 2D/3D

2016

ANDRÉ Michel, SAMARAS Zissis
Energy and Environment
(Research for Innovative Transports Set - Volume 1)

AUBRY Jean-François, BRINZEI Nicolae, MAZOUNI Mohammed-Habib
Systems Dependability Assessment: Benefits of Petri Net Models (Systems Dependability Assessment Set - Volume 1)

BLANQUART Corinne, CLAUSEN Uwe, JACOB Bernard
Towards Innovative Freight and Logistics (Research for Innovative Transports Set - Volume 2)

COHEN Simon, YANNIS George
Traffic Management (Research for Innovative Transports Set - Volume 3)

MARÉ Jean-Charles
Aerospace Actuators 1: Needs, Reliability and Hydraulic Power Solutions

REZG Nidhal, HAJEJ Zied, BOSCHIAN-CAMPANER Valerio
Production and Maintenance Optimization Problems: Logistic Constraints and Leasing Warranty Services

TORRENTI Jean-Michel, LA TORRE Francesca
Materials and Infrastructures 1 (Research for Innovative Transports Set - Volume 5A)
Materials and Infrastructures 2 (Research for Innovative Transports Set - Volume 5B)

WEBER Philippe, SIMON Christophe
Benefits of Bayesian Network Models
(Systems Dependability Assessment Set – Volume 2)

YANNIS George, COHEN Simon
Traffic Safety (Research for Innovative Transports Set - Volume 4)

2015

AUBRY Jean-François, BRINZEI Nicolae
Systems Dependability Assessment: Modeling with Graphs and Finite State Automata

BOULANGER Jean-Louis
CENELEC 50128 and IEC 62279 Standards

BRIFFAUT Jean-Pierre
E-Enabled Operations Management

MISSIKOFF Michele, CANDUCCI Massimo, MAIDEN Neil
Enterprise Innovation

2014

CHETTO Maryline
Real-time Systems Scheduling
Volume 1 – Fundamentals
Volume 2 – Focuses

DAVIM J. Paulo
Machinability of Advanced Materials

ESTAMPE Dominique
Supply Chain Performance and Evaluation Models

FAVRE Bernard
Introduction to Sustainable Transports

GAUTHIER Michaël, ANDREFF Nicolas, DOMBRE Etienne
Intracorporeal Robotics: From Milliscale to Nanoscale

MICOUIN Patrice
Model Based Systems Engineering: Fundamentals and Methods

MILLOT Patrick
Designing Human–Machine Cooperation Systems

NI Zhenjiang, PACORET Céline, BENOSMAN Ryad, RÉGNIER Stéphane
Haptic Feedback Teleoperation of Optical Tweezers

OUSTALOUP Alain
Diversity and Non-integer Differentiation for System Dynamics

REZG Nidhal, DELLAGI Sofien, KHATAD Abdelhakim
Joint Optimization of Maintenance and Production Policies

STEFANOIU Dan, BORNE Pierre, POPESCU Dumitru, FILIP Florin Gh.,
EL KAMEL Abdelkader
Optimization in Engineering Sciences: Metaheuristics, Stochastic Methods and Decision Support

2013

ALAZARD Daniel
Reverse Engineering in Control Design

ARIOUI Hichem, NEHAOUA Lamri
Driving Simulation

CHADLI Mohammed, COPPIER Hervé
Command-control for Real-time Systems

DAAFOUZ Jamal, TARBOURIECH Sophie, SIGALOTTI Mario
Hybrid Systems with Constraints

FEYEL Philippe
Loop-shaping Robust Control

FLAUS Jean-Marie
Risk Analysis: Socio-technical and Industrial Systems

FRIBOURG Laurent, SOULAT Romain
Control of Switching Systems by Invariance Analysis: Application to Power Electronics

GROSSARD Mathieu, REGNIER Stéphane, CHAILLET Nicolas
Flexible Robotics: Applications to Multiscale Manipulations

GRUNN Emmanuel, PHAM Anh Tuan
Modeling of Complex Systems: Application to Aeronautical Dynamics

HABIB Maki K., DAVIM J. Paulo
Interdisciplinary Mechatronics: Engineering Science and Research Development

HAMMADI Slim, KSOURI Mekki
Multimodal Transport Systems

JARBOUI Bassem, SIARRY Patrick, TEGHEM Jacques
Metaheuristics for Production Scheduling

KIRILLOV Oleg N., PELINOVSKY Dmitry E.
Nonlinear Physical Systems

LE Vu Tuan Hieu, STOICA Cristina, ALAMO Teodoro, CAMACHO Eduardo F., DUMUR Didier
Zonotopes: From Guaranteed State-estimation to Control

MACHADO Carolina, DAVIM J. Paulo
Management and Engineering Innovation

MORANA Joëlle
Sustainable Supply Chain Management

SANDOU Guillaume
Metaheuristic Optimization for the Design of Automatic Control Laws

STOICAN Florin, OLARU Sorin
Set-theoretic Fault Detection in Multisensor Systems

2012

AÏT-KADI Daoud, CHOUINARD Marc, MARCOTTE Suzanne, RIOPEL Diane
Sustainable Reverse Logistics Network: Engineering and Management

BORNE Pierre, POPESCU Dumitru, FILIP Florin G., STEFANOIU Dan
Optimization in Engineering Sciences: Exact Methods

CHADLI Mohammed, BORNE Pierre
Multiple Models Approach in Automation: Takagi-Sugeno Fuzzy Systems

DAVIM J. Paulo
Lasers in Manufacturing

DECLERCK Philippe
Discrete Event Systems in Dioid Algebra and Conventional Algebra

DOUMIATI Moustapha, CHARARA Ali, VICTORINO Alessandro, LECHNER Daniel
Vehicle Dynamics Estimation using Kalman Filtering: Experimental Validation

GUERRERO José A, LOZANO Rogelio
Flight Formation Control

HAMMADI Slim, KSOURI Mekki
Advanced Mobility and Transport Engineering

MAILLARD Pierre
Competitive Quality Strategies

MATTA Nada, VANDENBOOMGAERDE Yves, ARLAT Jean
Supervision and Safety of Complex Systems

POLER Raul *et al.*
Intelligent Non-hierarchical Manufacturing Networks

TROCCAZ Jocelyne
Medical Robotics

YALAOUI Alice, CHEHADE Hicham, YALAOUI Farouk, AMODEO Lionel
Optimization of Logistics

ZELM Martin *et al.*
Enterprise Interoperability –I-EASA12 Proceedings

2011

CANTOT Pascal, LUZEAUX Dominique
Simulation and Modeling of Systems of Systems

DAVIM J. Paulo
Mechatronics

DAVIM J. Paulo
Wood Machining

GROUS Ammar
Applied Metrology for Manufacturing Engineering

KOLSKI Christophe
Human–Computer Interactions in Transport

LUZEAUX Dominique, RUAULT Jean-René, WIPPLER Jean-Luc
Complex Systems and Systems of Systems Engineering

ZELM Martin, *et al.*
Enterprise Interoperability: IWEI2011 Proceedings

2010

BOTTA-GENOULAZ Valérie, CAMPAGNE Jean-Pierre, LLERENA Daniel, PELLEGRIN Claude
Supply Chain Performance / Collaboration, Alignement and Coordination

BOURLÈS Henri, GODFREY K.C. Kwan
Linear Systems

BOURRIÈRES Jean-Paul
Proceedings of CEISIE'09

CHAILLET Nicolas, REGNIER Stéphane
Microrobotics for Micromanipulation

DAVIM J. Paulo
Sustainable Manufacturing

GIORDANO Max, MATHIEU Luc, VILLENEUVE François
Product Life-Cycle Management / Geometric Variations

LOZANO Rogelio
Unmanned Aerial Vehicles / Embedded Control

LUZEAUX Dominique, RUAULT Jean-René
Systems of Systems

VILLENEUVE François, MATHIEU Luc
Geometric Tolerancing of Products

2009

DIAZ Michel
Petri Nets / Fundamental Models, Verification and Applications

OZEL Tugrul, DAVIM J. Paulo
Intelligent Machining

PITRAT Jacques
Artificial Beings

2008

ARTIGUES Christian, DEMASSEY Sophie, NERON Emmanuel
Resources–Constrained Project Scheduling

BILLAUT Jean-Charles, MOUKRIM Aziz, SANLAVILLE Eric
Flexibility and Robustness in Scheduling

DOCHAIN Denis
Bioprocess Control

LOPEZ Pierre, ROUBELLAT François
Production Scheduling

THIERRY Caroline, THOMAS André, BEL Gérard
Supply Chain Simulation and Management

2007

DE LARMINAT Philippe
Analysis and Control of Linear Systems

DOMBRE Etienne, KHALIL Wisama
Robot Manipulators

LAMNABHI Françoise *et al.*
Taming Heterogeneity and Complexity of Embedded Control

LIMNIOS Nikolaos
Fault Trees

2006

FRENCH COLLEGE OF METROLOGY
Metrology in Industry

NAJIM Kaddour
Control of Continuous Linear Systems

Printed and bound by CPI Group (UK) Ltd, Croydon, CR0 4YY
15/09/2026
14962362-0001